Statistical Significance and the PHC Curve

Hideki Toyoda

Statistical Significance and the PHC Curve

Springer

Hideki Toyoda
Department of Psychology
Waseda University
Shinjuku City, Tokyo, Japan

ISBN 978-981-97-7747-1 ISBN 978-981-97-7748-8 (eBook)
https://doi.org/10.1007/978-981-97-7748-8

This Springer imprint is published by the registered company Springer Nature Singapore Pte Ltd.
The registered company address is: 152 Beach Road, #21-01/04 Gateway East, Singapore 189721, Singapore

Preface

In March 2019, two impactful papers on statistics were published. One paper [1], “Moving to a World Beyond ‘$p < 0.05$’”, was featured in the scholarly journal *The American Statistician*, overseen by the American Statistical Association. The title of the first chapter is “Don’t Say ‘Statistically Significant’”, and it uses the imperative form to clearly forbid the use of significance testing. Continuing to use *p*-values is equivalent to driving a car that the leading manufacturer has recalled and banned from use.

Another paper [2], “Retire statistical significance”, was published in the prestigious scientific journal Nature. This commentary was endorsed by over 800 scientists, advocating for the statement, “We agree, and call for the entire concept of statistical significance to be abandoned.” Continuing to teach p-values is like driving a car which the leading consumer advocacy group has recommended against using.

The *p*-value is a probability, yet it is difficult to interpret intuitively. Numerical values used as analysis results are more user-friendly when they are intuitive and easily understood by statistical novices. Fortunately, such indicators are already in practical use and widely established in societal infrastructure. For example, they are implemented in the “posterior probability of being spam” filter used in our daily emails. Unlike *p*-values, this probability directly indicates the chance that an incoming email is spam. It is expressed in terms everyone can understand and feel, making it a tangible probability. How about adopting this methodology? Consider a study comparing the duration of hospital stays between treatments A and B. Previously, research conclusions were typically stated as: “There was a statistically significant difference at the 5% level in the average duration of hospital stays.” This phrasing is quite abstract. Instead, we suggest adapting to the specific context of the research objectives, setting flexible research questions, and responding accordingly. For example,

- The average duration of hospital stays for Group A is at least half a day shorter than for Group B.
- 71% of patients in Group A have shorter hospital stays than the average for Group B.

- Group A has an average hospital stay that is, on average, no more than 94% of that of Group B.

Wouldn't it be better to express conclusions in terms that are understandable not only to physicians but also to patients? And like an email filter, this expression would indicate the correct probability of the statement. Both analysts and readers can feel the effectiveness of this approach. This book explains the importance of using the **Probability that Hypothesis is Correct** (PHC), an intuitive measure that anyone can understand, as an alternative to the *p*-value.

All the source code and data in each chapter are freely available on this book's GitHub repository: https://github.com/TOYODA-Hideki/Statistical-Significance-and-the-PHC-Curve.

The computing environment under which this book was written was Windows 11 (64bit), R 4.2.2, Stan 2.31.0, CmdStanR 0.5.3.

The script that executes the contents of Chaps. 1–4 of the textbook is "post p01.R". The script that executes the contents of Chaps. 5–7 of the textbook is "post p02.R". The working directory in R language is "SSandPHC".

Tokyo, Japan
July 2024

Hideki Toyoda

Contents

Part I The Brink of Statistical Science

1 "Statistical Significance" Is Merely a Necessary Condition 3
1.1 Evidence of Predictive Ability Discovered!? 3
1.2 The Purpose of Testing 5
1.3 The Procedure of Testing 6
1.3.1 Sampling Distribution 7
1.3.2 *p*-Value 8
1.3.3 Significance Level 9
1.3.4 Rejection and Acceptance Regions 10
1.3.5 Interval Estimation Using Confidence Intervals 12
1.4 Test Results for "Feeling the Future" 12
1.5 Let's Try a "Telekinesis Experiment" 14
1.6 Summary 16
Column: Statement of the American Statistical Association 17

2 The Invisible Hand of God 19
2.1 Is There an Effect in Dieting Methods?/Crash Course in Statistical Terms 19
2.1.1 Histograms 20
2.1.2 Numerical Summary 21
2.1.3 Representative Value 22
2.1.4 Dispersion 23
2.2 Test of the Mean Difference of Paired Two Groups 23
2.3 Statistically Significant but Scientifically Meaningless 27
2.4 The Null Hypothesis Is Scientifically False 28
2.5 The *p*-Value Question in the National Medical Examination 29
2.6 The *p*-Value Is Probabilistic but Abstract 30
2.7 The Invisible Hand of God 32
2.7.1 Focusing Only on Profit Leads to Appropriate Pricing 32

2.7.2 Focusing Only on Statistical Significance Leads to a High and Stable Proportion of Meaningless Papers in Journals 33
2.8 Conclusion 34

3 Tiger at the Front Gate, Wolf at the Back Gate 37
3.1 Efficacy of Antibacterial Drug A 37
3.2 Test for Mean Difference Between Two Independent Groups 39
3.2.1 Difference in Average Hospitalization Days Required for Statistical Significance 41
3.3 Two Types of Errors Associated with Decision-Making 42
3.3.1 Reasons Not to Further Reduce the Risk Level 42
3.3.2 Power of a Test 44
3.4 The Null Hypothesis Cannot Be Adopted 44
3.4.1 Significance Means Finding an Alibi 45
3.4.2 Attitude Reservation if No Alibi Is Found 46
3.4.3 The Null Hypothesis Is False Before Data Collection 46
3.5 Tiger at the Front Gate, Wolf at the Back Gate 47
3.6 Power Analysis 47
3.6.1 Determining n for Tests of Proportions 48
3.6.2 Determining the Sample Size n for the Test of Mean Differences Between Two Independent Groups 49
3.6.3 Determining n for the Test of Mean Differences Between Two Related Groups 50
3.7 Limitations of Power Analysis 51
3.8 Summary 53
Column: The Multiple Comparisons Problem 54

4 The Zombie Problem 59
4.1 Testing Every Time 59
4.2 The Sample Distribution Relies on Unrecorded Intent 62
4.2.1 Intent 1: Fixing n and Observing x 62
4.2.2 Intent 2: Fixing x and Observing n 63
4.2.3 Intent 3: Fixing Research Resources, Observing n, and Then x 64
4.2.4 Intent 4: Increasing n While Monitoring x and p-Values 66
4.3 The Zombie Problem in Significance Testing 67
4.3.1 Philosophical Zombies and Zombie-Like Tests (Definition of Terms) 68
4.3.2 Eradicating Zombies 68
4.4 Confidence Intervals Do Not Save Significance Testing 69
4.5 Pre-registration System 70
4.5.1 Benefits of Pre-registration 71

4.5.2 Problems with Pre-registration in Significance Testing ... 71
4.6 Conclusion ... 73
Interlude: Problems Revealed in Part I ... 74
Interlude: Issues Revealed in Part I ... 76

Part II Saving the Field of Statistics!

5 Bayes' Theorem—"The Probability of a Research Hypothesis Being Correct" ... 81
5.1 The Probability of Being Spam ... 81
5.1.1 Joint Probability and Conditional Probability ... 81
5.1.2 Bayes' Theorem ... 83
5.1.3 Probability of Not Being Spam ... 84
5.1.4 Transformation of Bayes' Theorem 1 ... 84
5.1.5 Transformation of Bayes' Theorem 2 ... 85
5.1.6 Bayesian Updating ... 85
5.1.7 Comparison with p-Values ... 87
5.2 Bayes' Theorem for Distributions ... 88
5.3 Likelihood Principle ... 88
5.3.1 Likelihood ... 89
5.3.2 Likelihood Does Not Depend on Intent ... 89
5.3.3 Prior and Posterior Distributions ... 90
5.4 Estimating the Population Proportion for "Premonition of the Future" ... 91
5.4.1 Probability that a Research Hypothesis Is True ... 92
5.4.2 Peer Review with Metrics Convincing to Those with Domain Knowledge ... 93
5.4.3 Determining the Baseline c ... 94
5.4.4 It Is Not Necessary to Set a Baseline c/PHC Curve ... 94
5.4.5 Actively Presenting Claims Equivalent to the Null Hypothesis ... 95
5.4.6 Clarification of Hypothesis Truth as the Number of Observations Increases ... 96
5.4.7 The Binary Judgment of Significance or Non-significance Involves Great Social Disadvantages ... 97
5.5 Conclusion ... 98

6 Concluding Remarks with Sincerity ... 99
6.1 Inference of the Difference Between Two Independent Groups ... 99
6.1.1 Modeling the Data Generation Process ... 99
6.1.2 Bayes' Theorem ... 100

6.1.3 Likelihood ... 101
6.2 Breaking Free from Zombies ... 101
6.3 Prior Distribution ... 102
6.3.1 Uniform Distribution ... 102
6.3.2 Determining the Parameters of the Uniform Distribution ... 103
6.3.3 The Upper and Lower Limits of the Uniform Distribution Can Be Determined Loosely ... 103
6.3.4 Unique Properties of the Uniform Distribution as a Prior ... 104
6.4 Inference of the Difference Between Two Independent Groups ... 105
6.4.1 EAP, post.sd, Credible Intervals ... 106
6.5 Generative Quantities ... 107
6.6 Posterior Distribution of the Difference in Population Means ... 107
6.6.1 Probability that the Hypothesis of a Difference Is Correct ... 108
6.6.2 Probability that the Hypothesis of No Difference Is Correct ... 110
6.7 Standardized Mean Difference ... 111
6.7.1 Power Analysis May Not Always Be Valid ... 111
6.7.2 Posterior Distribution of the Standardized Mean Difference ... 112
6.7.3 Probability that the Hypothesis $\delta_{\text{con}} > c$ Is Correct ... 113
6.8 Non-overlap Degree ... 114
6.9 Threshold Exceedance Rate ... 115
6.10 Inference of Ratio ... 117
Interlude: Resolved Issues from Part I ... 119

Part III Education and Malice

7 Seligman's Dogs ... 125
7.1 The Diet Problem Revisited ... 125
7.1.1 Test for the Difference Between Two Corresponding Groups ... 125
7.1.2 Research Questions ... 127
7.1.3 In the Case of a Learner Who Memorized Test Statistics Without Understanding ... 128
7.2 Likelihood Modeling ... 129
7.2.1 Freedom from the Multiplicity Problem ... 130

7.2.2 In the Case of a Learner Who Memorized that the Posterior Distribution Can Be Obtained by MCMC Without Understanding 131
7.3 PHC Makes the Meaning of Analytical Results More Tangible 133
7.4 Properties of PHC Curves 134
7.5 The Future of Statistical Education 136
7.5.1 Top-Down Approach and Learned Helplessness 137
7.5.2 Top-Down Approach and Sense of Control/Efficacy 138
7.5.3 Fostering a Sense of Competence and Control in Modeling the Data Generation Process 138

8 Tampering and Cover-Up 141
8.1 Tampering 141
8.2 Cover-Up 142
8.2.1 Cover-Up by Thinning Out Variables 142
8.2.2 Cover-Up by Selecting Periods 143
8.3 HARKing and Cover-Up 143
8.4 Rethinking the Pre-registration System 144
8.4.1 Review of Pre-registration in the Absence of Cover-Up 144
8.4.2 Differences Between *p*-Values and PHC in Pre-registration 145
8.4.3 Publication Bias and Peer-Reviewed Pre-registration System 146
8.4.4 PHC and Pre-registration 147
8.5 Saving Dying Statistics! 147

Q & A 149

References 155

Part I
The Brink of Statistical Science

Chapter 1
"Statistical Significance" Is Merely a Necessary Condition

Criteria for Peer Review Unrelated to Scholarly Value.

1.1 Evidence of Predictive Ability Discovered!?

This intriguing story begins with a peculiar paper. Dr. Daryl Bem, an emeritus professor at Cornell University and a renowned social psychologist, is well-known for his EBE theory (a theory on how feelings of sexual orientation diversity transform into erotic love), which is featured in many textbooks.

It was in 2011. In JPSP, the most prestigious journal in social psychology, Bem suddenly published a paper titled "Feeling the Future" [5] (see Fig. 1.1). The paper, with its somewhat strange and romantic title, contained numerous experiments and their analyses, but the first experiment was particularly noteworthy. The experiment, which involved nude photographs, went roughly as follows:

> On a computer screen, two curtains are displayed side by side. Participants are instructed: "Behind one of the curtains is a nude photo of the opposite sex, and behind the other, there is none. Please click on the curtain you think has the photo." After the participant clicks on the curtain they feel has the nude photo, a random number generator sets the photo behind only one curtain. The curtain is then opened to check if the photo is there. This counts as one trial. Out of 1560 trials, 829 correctly guessed the location of the nude photo.

H. Toyoda, *Statistical Significance and the PHC Curve*,
https://doi.org/10.1007/978-981-97-7748-8_1

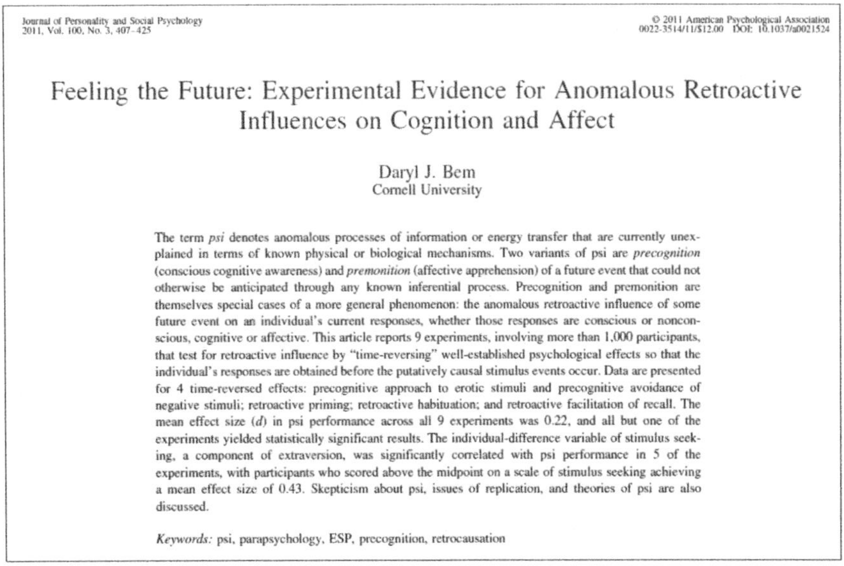

Journal of Personality and Social Psychology
2011, Vol. 100, No. 3, 407–425

© 2011 American Psychological Association
0022-3514/11/$12.00 DOI: 10.1037/a0021524

Feeling the Future: Experimental Evidence for Anomalous Retroactive Influences on Cognition and Affect

Daryl J. Bem
Cornell University

The term *psi* denotes anomalous processes of information or energy transfer that are currently unexplained in terms of known physical or biological mechanisms. Two variants of psi are *precognition* (conscious cognitive awareness) and *premonition* (affective apprehension) of a future event that could not otherwise be anticipated through any known inferential process. Precognition and premonition are themselves special cases of a more general phenomenon: the anomalous retroactive influence of some future event on an individual's current responses, whether those responses are conscious or nonconscious, cognitive or affective. This article reports 9 experiments, involving more than 1,000 participants, that test for retroactive influence by "time-reversing" well-established psychological effects so that the individual's responses are obtained before the putatively causal stimulus events occur. Data are presented for 4 time-reversed effects: precognitive approach to erotic stimuli and precognitive avoidance of negative stimuli; retroactive priming; retroactive habituation; and retroactive facilitation of recall. The mean effect size (*d*) in psi performance across all 9 experiments was 0.22, and all but one of the experiments yielded statistically significant results. The individual-difference variable of stimulus seeking, a component of extraversion, was significantly correlated with psi performance in 5 of the experiments, with participants who scored above the midpoint on a scale of stimulus seeking achieving a mean effect size of 0.43. Skepticism about psi, issues of replication, and theories of psi are also discussed.

Keywords: psi, parapsychology, ESP, precognition, retrocausation

Fig. 1.1 Cover of Bem's paper titled "Feeling the Future"

Bem analyzed this data using a statistical method called significance testing, and applied for publication in JPSP based on the statistical significance of "Feeling the Future". In academic journals, papers are usually subjected to a review process, known as peer review, to determine their scholarly value. Many academic papers that use experiments and surveys have been published based on the results of significance tests in statistics being deemed significant.

In significance testing, a probability known as the p-value is typically used, and a result is considered "statistically significant" when this value falls below 5%. In many fields of research, regardless of the analysis purpose or subject matter, the criterion of "whether the p-value is less than 0.05" has been used as a standard for value judgment.

The p-value of Bem's precognition experiment was 1.3%, satisfying the condition $p < 0.05$. The reviewers at JPSP were in a dilemma. Publishing a paper claiming the existence of precognition could tarnish the reputation of a prestigious academic journal. However, rejecting Bem's paper solely because it seemed implausible, despite meeting the $p < 0.05$ criterion, would not be fair or impartial. After much deliberation, the reviewers accepted Bem's argument, and "Feeling the Future" was eventually published in JPSP.

As expected, Bem's paper caused a significant stir and faced intense criticism from various quarters. Of course, psychologists immediately began replication experiments, but no one could reproduce the results. Although statistically significant, the paper was psychologically meaningless.

1.2 The Purpose of Testing

The reason "Feeling the Future" was published in a prestigious journal of social psychology was that it met the condition of $p < 0.05$ required by the significance test. But what exactly is a significance test? There are many types of significance tests. Here, we will take the binomial test used by Bem as an example and concretely explain the general idea behind significance testing.

In statistics, an attempt that can result in different outcomes probabilistically is called a **trial**. Suppose we conduct ten trials to predict the location of a nude photo. The possible outcomes observed are called **events**. Hits and misses are events. For example, let's say the location was correctly predicted seven times. The number of trials is called the **number of trials** n, and the number of hits is called the **number of successes** x. The ratio of the number of successes to the number of trials is called the **sample proportion**, and is calculated as

$$\text{Sample Proportion} = \frac{\text{Number of Successes}}{\text{Number of Trials}} = \frac{7}{10} = 0.7. \qquad (1.1)$$

The sample proportion is a ratio calculated from the data.

Some people might consider a 70% success rate in predicting the location of a nude photo as evidence of precognitive ability. But does this sample proportion indicate the existence of precognitive ability? No, it does not necessarily indicate that.

Because even flipping a fair, unbiased coin, it's not unusual to get heads 7 out of 10 times. In trials like a coin toss, where there are only two possible outcomes (heads or tails), the probability of getting heads is represented by a probability distribution called the **binomial distribution** (Fig. 1.2). A **probability distribution** shows how likely different values are to be observed. A probability distribution is sometimes simply referred to as a distribution. The binomial distribution is one type of probability distribution.

According to Fig. 1.2, the probability of getting heads 7 out of 10 times in a coin toss is 11.7%. This can be observed more than once in every 10 trials if repeated. This time, the location of the nude photo was correctly predicted 7 times, but like a coin toss, it might not be so accurate next time. Therefore, just considering the sample proportion leads to endless debates like "It's high enough, so there must be precognitive ability!" or "No, it's just a coincidence, like a coin toss."

The most academically important thing is the reproducibility of the phenomenon. What matters is that the average ratio in an infinite number of replication experiments under the same conditions is sufficiently high. If that is demonstrated, it becomes clear to anyone that precognitive ability exists.

The average ratio mentioned here is different from the sample proportion calculated from the current data, so to distinguish it, let's call it the **population proportion** π (pi). Academically, it's the value of the population proportion, not the sample proportion, that is of interest.

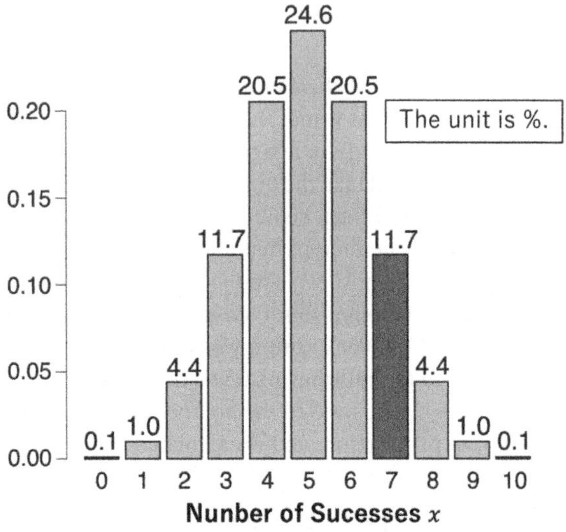

Fig. 1.2 Binomial distribution with trial number $n = 10$, population proportion $\pi = 0.5$

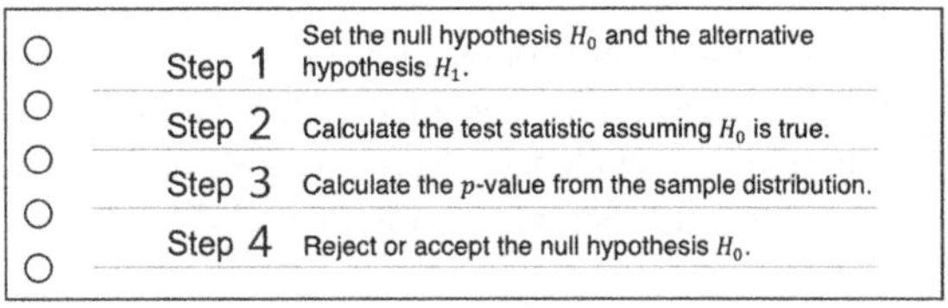

Fig. 1.3 Flow of significance testing

In a coin toss, the population proportion of getting heads is $\pi = 0.5$. To "demonstrate the existence of supernatural abilities," at the very least, it must be shown that the population proportion of correctly predicting the location of the nude photo is not $\pi = 0.5$. It's a necessary condition, but not more than that.

1.3 The Procedure of Testing

Even though academic interest lies in the population proportion π, not the sample proportion, it's practically impossible to conduct an infinite number of replication experiments. Hence, the introduction of the binomial test, a type of significance test. The **binomial test** is a method to make a certain judgment about the population proportion (not the sample proportion) from a single experiment's data. The procedure is shown in Fig. 1.3. This procedure is common to many **significance tests**.

The "Flow of Significance Testing" in Fig. 1.3 is full of technical terms, so let's explain them one by one carefully.

In **step 1**, we set the null hypothesis H_0 and the alternative hypothesis H_1.

Specifically, the null hypothesis is

$$H_0 : \pi = 0.5. \tag{1.2}$$

This **null hypothesis** means "The population proportion of correctly predicting the location of the nude photo is 0.5, just a coincidence." It's the hypothesis that researcher Bem wants to nullify (reject). The negation of the null hypothesis is a necessary condition for the existence of precognitive ability.

The **alternative hypothesis** is the negation of the null hypothesis. We set $H_1 : \pi \neq 0.5$, meaning "It's not just a coincidence." The alternative hypothesis is the hypothesis that the researcher ultimately wants to accept.

1.3.1 Sampling Distribution

In **step 2**, we calculate the test statistic assuming H_0 is true.

A **statistic** is a function of the data

$$\text{Statistic} = f(\text{Data}). \tag{1.3}$$

The number of successes x and the sample proportion, which have already been mentioned, are calculated from the data, so they are statistics.

A **test statistic** is a statistic used in a test. In a binomial test, either the number of successes x or the sample proportion can be used as the test statistic. In the example of Eq. (1.1), the number of successes was $x = 7$, and the sample proportion was 0.7. Using either as the test statistic will yield the same test result.

In **step 3**, we calculate the p-value from the sampling distribution.

The **sampling distribution** is the distribution of a statistic (e.g., the number of successes x). Assume that the null hypothesis (1.2) is true and imagine conducting an infinite number of replication experiments with a trial number n. Further, imagine checking the number of successes x in each experiment. The distribution of the number of successes in this case is called the sampling distribution of x. Since the experimental results are either success or failure, the sampling distribution of the number of successes x is a binomial distribution.

We don't actually conduct an infinite number of replication experiments. Testing is applied to the data from a single experiment. Therefore, the sampling distribution is a theoretical concept and is not actually observed. It's an imaginary, non-existent distribution. Be careful not to confuse the sampling distribution with the distribution of the sample (the distribution of the data).

The previously mentioned Fig. 1.2 is the binomial distribution of x when $n = 10$. The **binomial distribution** is generally denoted as

$$\text{Binomial Distribution}_n(x|\pi). \tag{1.4}$$

The inside of the brackets is read as "x given pi." The **vertical bar** | means "given," indicating the probability distribution of the number of successes x under the condition of a given population proportion π. Instead of saying "The probability of getting heads 7 out of 10 times in a coin toss is 11.7%," we can also write

$$\text{Binomial Distribution}_{n=10}(x = 7|\pi = 0.5) = 0.117. \tag{1.5}$$

1.3.2 p-Value

After calculating the test statistic x, we evaluate its probability. This is where the p-value comes in. **The p-value is the probability, under the assumption that the null hypothesis is true, that the test statistic in the sampling distribution takes on a value as extreme or more extreme than the value calculated from the data.**

Note that it's not the probability of the test statistic exactly matching the value calculated from the data. To reiterate, it's not the probability of observing the exact number of successes, 7, in this case. Under the condition that the null hypothesis is true, the p-value is the probability of observing an event as extreme or more extreme than the observed number of successes, 7.

Considering 5 successes as a baseline, an extreme value of 7 or more also includes the region of 3 or less. Specifically, the p-value is the probability of observing an event where x is one of $\{0, 1, 2, 3, 7, 8, 9, 10\}$. Picking the corresponding probabilities from Fig. 1.2, we get

$$(11.7 + 4.4 + 1.0 + 0.1) \times 2 = 34.4\%. \tag{1.6}$$

Thus, $p = 0.344$.

Why isn't the p-value the probability of the test statistic exactly matching the value calculated from the data? Because it's difficult to assess whether that probability is high or low.

Let's consider an experimental data set with 100 trials and 70 hits. The sample proportion remains the same at 0.7. However, evaluating the probability using the binomial distribution gives

$$\text{Binomial Distribution}_{n=100}(x = 70|\pi = 0.5) = 0.00002. \tag{1.7}$$

The probability of exactly observing $x = 70$ is just 0.00002.

With 100 trials, there are 101 possible values for x, so each one has a small probability. The probability of observing the test statistic exactly as calculated from the data strongly depends on the size of n.

Therefore, as a probability to judge whether an event is likely or unlikely, it's not suitable. In other words, since the number of possible outcomes of the trial results is originally different, like 11 and 101, the probability of observing the sample proportion of 0.7 itself is not an appropriate metric for evaluation.

For 100 trials, the p-value is the probability of observing an event as extreme or more extreme than the observed 70 hits. With a binomial distribution of $n = 100$, there are 101 bars, and summing the probabilities of the relevant bars gives a p-value of

$$p(0 \leq x \leq 30, \text{ or } 70 \leq x \leq 100) = 0.000079. \tag{1.8}$$

Here, $p(\,)$ denotes the probability of the event inside the brackets occurring. It's also fine to say that the probability of observing a sample proportion as extreme or more extreme than 0.7 is 0.000079.

In the case of 10 trials, the probability of the sample proportion being as extreme or more extreme than 0.7 was 34.4%. In contrast, for 100 trials, the probability of the same event occurring is only 0.0079%, which is extremely small. It means that even when calculating the probability of an event more extreme than the observed value, the p-value is smaller for 100 trials than for 10 trials, given the same sample proportion.

In other words, if you toss a coin 10 times, it's quite common for the sample proportion to be as extreme or more extreme than 0.7. However, if you toss it 100 times, it's almost impossible for the sample proportion to be that extreme.

Using the p-value, which is the probability of an event more extreme than the calculated test statistic, allows us to fairly reflect such realities and correctly compare the probabilities.

1.3.3 Significance Level

In **step 4**, we either reject or accept the null hypothesis H_0.

In significance testing, if a p-value is small and an unlikely fact is observed, we conclude that the assumption that the null hypothesis is true is incorrect, and instead, "the null hypothesis is false." This is called **rejecting the null hypothesis**. In this case, we **accept the alternative hypothesis**.

However, the term "unlikely" is literary and subjective. Therefore, we define it in advance in terms of probability before looking at the results. The pre-determined probability of unlikeliness is called the **significance level**, denoted as α. Generally, 0.05 is used as the significance level.

There is no specific reason why the significance level is 5%. It is frequently used across all research fields, regardless of the analysis purpose, application subject, or importance of the results, with $\alpha = 0.05$ being common.

For an experiment with $n = 10$ trials and a sample proportion of 0.7, $p = 0.344 > 0.05$. An unlikely value under the assumption that the null hypothesis is true has not been observed. In this case, we conclude that there is no evidence that "the null hypothesis is false." This is called **accepting the null hypothesis**. In this case, we **reject the alternative hypothesis**. Essentially, it means that hitting 7 out of 10 times can "occur even in a coin toss."

For an experiment with $n = 100$ trials and a sample proportion of 0.7, $p = 0.000079 < 0.05$. We conclude that the assumption that the null hypothesis is true is incorrect, and instead, "the null hypothesis is false." We reject the null hypothesis and accept the alternative hypothesis. This state is referred to as "**significant difference**" or simply "**significant**." Essentially, it means "Hitting 70 out of 100 times is different from a trial like a coin toss."

For the null hypothesis of the population proportion $\pi = 0.5$, a sample proportion of 0.7 is not statistically significant at the 5% level for $n = 10$. However, it is statistically significant at the 5% level for $n = 100$.

1.3.4 Rejection and Acceptance Regions

In the case of $n = 10$ trials, how extreme does the sample proportion need to be for the p-value to be within 5%? Figure 1.4 shows the sample distribution of the sample proportion when $n = 10$ and $\pi = 0.5$. This figure is exactly the same as the binomial distribution in Fig. 1.2, but the horizontal axis has been changed from the number of successes x to the sample proportion. This is to facilitate comparison with the case of $n = 100$ trials later.

If the sample proportion is 1.0 or 0.0, the p-value is 0.2% ($= 0.1 + 0.1$), so it is significant at $p < 0.05$. If the sample proportion is 0.9 or 0.1, the p-value is 2.2% ($= (1.0 + 0.1) \times 2$), so it is significant at $p < 0.05$. However, if the sample proportion is 0.8 or 0.2, the p-value is 11.0% ($= (4.4 + 1.0 + 0.1) \times 2$), so it is no longer significant at $p > 0.05$. As it gets closer to 0.5, it becomes insignificant.

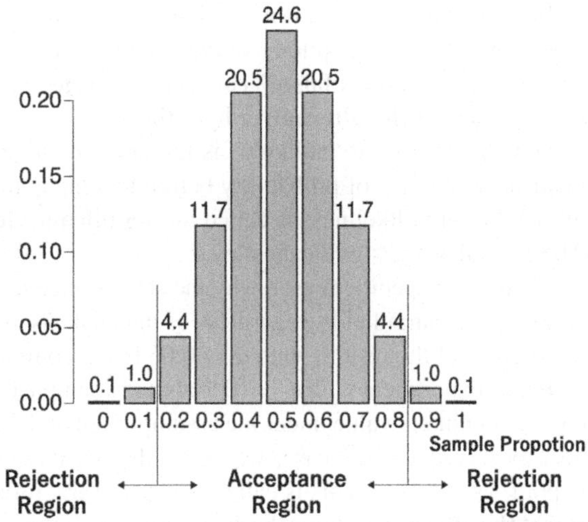

Fig. 1.4 Sample distribution of sample proportion and regions of rejection and acceptance (trial number $n = 10$)

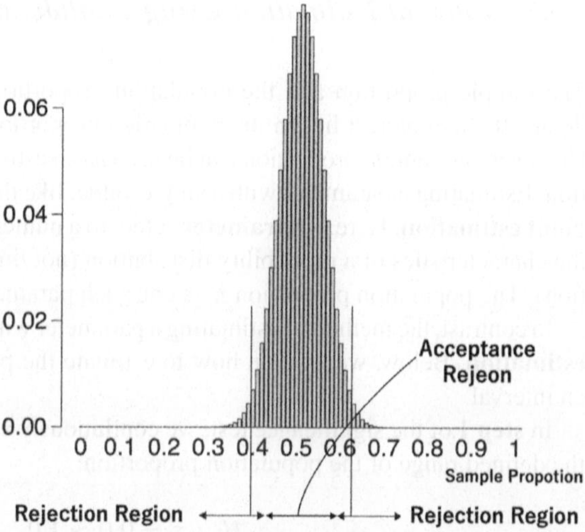

Fig. 1.5 Sample distribution of sample proportion and regions of rejection and acceptance (trial number $n = 100$)

When statistically significant, we reject the null hypothesis, and when not significant, we accept the null hypothesis. Therefore, in the sample distribution, the region of the test statistic that leads to a significant test result is called the **rejection region**. {0.0, 0.1, 0.9, 1.0} are in the rejection region. The region of the test statistic that does not lead to a significant test result in the sample distribution is called the **acceptance region**. {0.2, 0.3, 0.4, 0.5, 0.6, 0.7, 0.8} are in the acceptance region.

Figure 1.5 shows the sample distribution of the sample proportion when the number of trials is $n = 100$. The horizontal axis is exactly the same as in Fig. 1.4. While there were 11 bars for $n = 10$ trials, there are 101 bars for $n = 100$ trials.

It may seem like there are no bars at the extreme ends, but this is only because they are too low to be printed; bars do exist from 0.00 to 1.00.

Compared to $n = 10$, the most notable feature of the sample distribution for $n = 100$ is that the shape of the distribution is tightly compressed towards the center.

The rejection region for the sample proportion is

$$\{0.00 \leq \text{Sample Proportion} \leq 0.39, \text{ and } 0.60 \leq \text{Sample Proportion} \leq 1.00\}.$$

If you predict correctly 60 times or more, you can write in your paper that there is a significant difference at the 5% level, rejecting the null hypothesis. The acceptance region is

$$\{0.40 \leq \text{Sample Proportion} \leq 0.59\}.$$

For $n = 10$, the width of the acceptance region was 0.6 (= 0.8 − 0.2). Compared to that, for $n = 100$, the width of the acceptance region is 0.19 (= 0.59 − 0.40). Generally, if the significance level and sample proportion remain the same, the acceptance region narrows as n increases. Conversely, the rejection region expands.

1.3.5 Interval Estimation Using Confidence Intervals

The sample proportion and the population proportion are distinct concepts. Academically, the interest lies in the population proportion, not the sample proportion. However, the sample proportion can be used as an estimate of the population proportion. Estimating a parameter with a single value, like the sample proportion, is called **point estimation**. Here, a **parameter** refers to a numerical indicator that determines the characteristics of a probability distribution (not limited to the binomial distribution). The population proportion π is one such parameter.

In contrast, the method of estimating a parameter with an interval is called **interval estimation**. Below, we explain how to estimate the population proportion π using an interval.

In **step 1** of the significance test, we continuously vary the null hypothesis within the defined range of the population proportion:

$$H_0 : \pi = 0.0 \sim 1.0. \tag{1.9}$$

We conduct tests at the significance level α for each, and determine the lower limit π_L and the upper limit π_H where the null hypothesis is not rejected. The interval (π_L, π_H) is called the $((1 - \alpha) \times 100)\%$ **confidence interval**.

For $\alpha = 0.05$, it's called the **95% confidence interval**, and for $\alpha = 0.01$, it's the 99% confidence interval. In principle, any value of the significance level α is acceptable, but traditionally, the 95% confidence interval is most frequently used.

For $n = 10$ trials and $x = 7$ successes, the 95% confidence interval is [0.348, 0.933], with a width of 0.585. For $n = 100$ trials and $x = 70$ successes, the 95% confidence interval is [0.600, 0.788], with a width of 0.188. It's significantly narrower. If the sample proportion and significance level remain the same, the larger the number of trials, the narrower the confidence interval becomes.

1.4 Test Results for "Feeling the Future"

Using the mechanism of the binomial test, let's reanalyze the significance test of Bem's "Feeling the Future."[1] The location of the nude photos was correctly predicted 829 times out of 1560 trials.[2]

[1] In "Feeling the Future," a test statistic approximated by the normal distribution is presented. Here, instead of using the approximation, let's perform the binomial test we just learned about. This approach will yield more precise results.

[2] The paper states that the sample proportion is 0.531. From this, we can infer that the number of hits is $x = 1560 \times 0.531 \simeq 829$ or 828 times. Here, to favor the claim of precognitive ability, we assume the number of hits is $x = 829$.

In **step 1**, we set the null hypothesis H_0 and the alternative hypothesis H_1.

$$H_0 : \pi = 0.5, \quad H_1 : \pi \neq 0.5.$$

In **step 2**, we calculate the test statistic assuming H_0 is true.

In the binomial test, either the number of successes or the sample proportion can be used as the test statistic, yielding the same result. The number of successes is $x = 829$, and the sample proportion is 0.531 (= 829/1560). Half of the number of trials is 780 (= 1560/2), so there are 49 more hits than half.

In **step 3**, we calculate the p-value from the sampling distribution.

The p-value is the probability of the event "the number of successes being 829 or more, or 731 (= 780 − 49) or less" occurring. Assuming the null hypothesis is true ($\pi = 0.5$), we evaluate it using the binomial distribution for $n = 1560$ trials. Summing up the probabilities of the relevant number of successes, the p-value is 1.4%.

In **step 4**, we either reject or accept the null hypothesis H_0.

If the p-value is smaller than the significance level, we conclude that "it's not that the null hypothesis is correct and an unlikely event has occurred, but rather the null hypothesis was wrong." This is the rejection of the null hypothesis.

A common significance level used is 5%. The result of the binomial test is $p = 0.014 < 0.05$, which is significant. Therefore, Bem claimed the existence of precognitive ability. The reviewers of the academic journal JPSP acknowledged his claim, and the paper was published. However, immediate replication experiments did not reproduce the results.

Figure 1.6 shows the sample distribution for the binomial distribution with $n =$ 1560 trials. Both the number of successes and the sample proportion are marked on the horizontal axis. Except for a small central part, it may seem like there are no bars at the ends, but this is only because they are too low to be printed; there are indeed 1561 bars from 0 to 1560.

Fig. 1.6 Sample distribution for trial number $n = 1560$

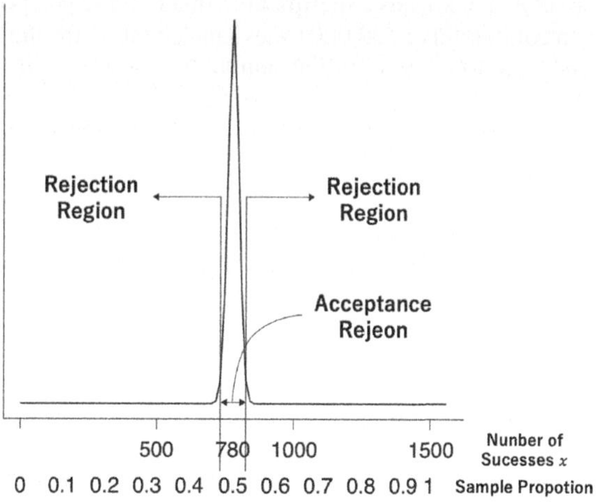

The rejection region for the number of successes when $n = 1560$ is

$$\{0 \le x \le 741, \text{ and } 819 \le x \le 1560\}.$$

The central small range is the acceptance region. The rejection region for the sample proportion is

$$\{0.000 \le \text{Sample Proportion} \le 0.475, \text{ and } 0.525 \le \text{Sample Proportion} \le 1.000\}.$$

If more than 52.5% are predicted correctly, Bem could have written in his paper that the null hypothesis is rejected at the 5% level and is "significant."

Of course, to "demonstrate the existence of supernatural abilities," at the very least, it must be shown that π is not 0.5. However, this is only a necessary condition. With 53.1%, no one would actually think there is precognitive ability. It is incorrect to use

$$H_0 : \pi = 0.5 \quad \text{with} \quad p < 0.05$$

as a criterion for judgment in all research fields, regardless of the analysis purpose or application subject.

The 95% confidence interval is [0.506, 0.556]. Since the test result is significant, the confidence interval does not include $\pi = 0.5$, set in the null hypothesis. However, this is not by chance, but a property derived from the definition of the confidence interval.

1.5 Let's Try a "Telekinesis Experiment"

What would happen in a binomial test if, for instance, the number of trials was 100,000? Predicting the location of a photo 100,000 times is challenging, so how about a "telekinesis experiment"? Imagine focusing on "heads" and flipping a one-yen coin a total of 100,000 times[3] and counting the number of times it lands on heads. The rejection region for the sample proportion in this binomial test would be

$$\{0.0000 \le \text{Sample Proportion} \le 0.4969, \text{ and } 0.5031 \le \text{Sample Proportion} \le 1.0000\}.$$

If more than 50.3% are heads, you can write in your paper that it is "significant at $p < 0.05$." For journals like JPSP, which require a p-value below 5% for publication, someone like Bem could claim "telekinesis exists." The sample distribution

[3] "What if you exchange a 100-yen coin for 100 one-yen coins, strongly focus on 'heads' and toss all 100 coins at once. Repeat this 10 times and count how many times heads appear out of 1000 coins." If you assign this task to students in a class of 100, you can conduct a "telekinesis experiment" with $n = 100{,}000$ trials in one class session.

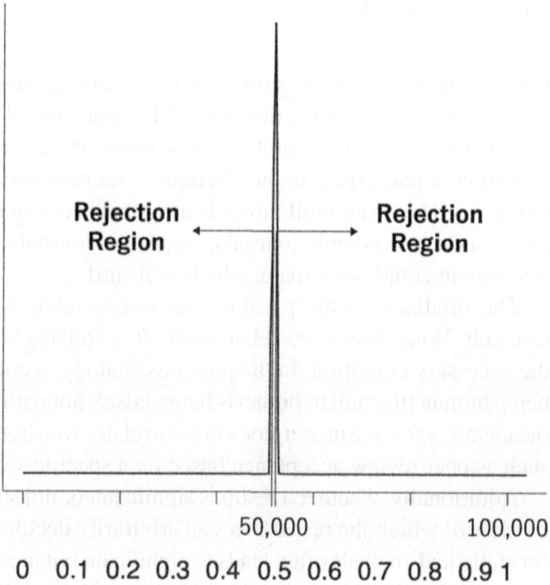

Fig. 1.7 Sample distribution for trial number $n = 100{,}000$

is covered in the rejection region like a needle, with the acceptance region being as small as the eye of the needle.

Figures 1.4, 1.5, 1.6, and 1.7 all have the sample proportion on the horizontal axis, ranging from 0 to 1. In this sense, the shapes of all four figures are comparable. When viewed in sequence, it's observed that with the increase in n, the rejection region rapidly expands, and the acceptance region narrows.

For $n = 1560$, the acceptance region is 0.5 ± 0.025. When the number of trials increases to $n = 100{,}000$, the acceptance region becomes 0.5 ± 0.0031. Generally, as n increases, the acceptance region narrows around 0.5 as much as possible, while the rejection region expands.

Assuming the sample proportion is the same as Bem's data at 0.531, it means heads appeared 53,100 times. In this case, the p-value, compared to the significance level of 0.05, is

$$p = 2.2 \times 10^{-16} = 0.00000000000000022, \tag{1.10}$$

an incredibly small probability.[4] A smaller p-value allows us to reject the null hypothesis with more confidence, but that's the extent of its benefit. When the p-value is much smaller than 0.05, it is described as **highly significant**. In the above equation, it is indeed highly significant. However, common sense dictates that no one would claim "telekinesis exists" with 0.531.

[4] "Even if we make the criterion stricter, say changing it to 0.005, meaningless papers can still be produced in abundance," as shown by Eq. (1.10).

1.6 Summary

In research aimed at finding differences, "rejecting the null hypothesis $H_0 : \pi = 0.5$" does not mean a "scientific discovery." The rejection of the null hypothesis is merely a very modest necessary condition for scientific discovery. Being human is a necessary condition to participate in the Olympics, but not every human can participate in the Olympics. However, until now, clearing this necessary condition was a criterion for publication in academic journals. Academic journals were like an Olympics where any human could participate, which is absurd.

The smallness of the p-value does not correlate with the academic value of the research. What does it correlate with? It correlates with the confidence in clearing the necessary condition. In the previous analogy, it correlates with the confidence in being human (the null hypothesis being false), not with the performance of the athlete (academic value). Since it does not correlate, we should not make binary decisions such as peer review acceptance based on a specific value like 0.05.

Additionally, whether a result is significant or not strongly depends on the number of trials n, which the researcher can arbitrarily decide. Therefore, using the p-value for statistical analysis often leads to significant but meaningless results. Bem's "Feeling the Future" controversy is a symbolic incident of this. It's an example of the harm caused by the mechanical decision-making based on the $p < 0.05$ criterion.

However, this became a major issue because it was a sensational research topic like an experiment on precognitive ability using nude photos. If it had been a "modest and serious research topic," it would have been quietly published as if nothing had happened. This negative impact has been like a body blow, leading to a large number of studies being reported without reproducibility.

People in general society believe that each study published in academic journals contributes to the advancement of knowledge. However, as mentioned in the 2015 Science paper [3] introduced in the "Introduction," only 25% of social psychology studies could reproduce their significance. This happened in papers published in top-tier journals. In subsequent academic journals, this rate is expected to be even lower. The negative impact of significance testing on science is immeasurable. The mystery of why this rate is so low will be unraveled in Chap. 4.

Does this mean that one in four papers published in top-tier journals is beneficial? No, it does not. The fact that reproducibility has been confirmed from the perspective of significance testing only means that there is a high probability that 25% of the papers have cleared the necessary condition. These 25% of papers are further classified into (a) those that are statistically significant and scientifically meaningful, and (b) those that are statistically significant but scientifically meaningless. The 25% consists of top athletes and mere humans. What is the ratio of this composition ($a : b = athletes : humans$)? This issue will be examined in detail in Chap. 2.

Column: Statement of the American Statistical Association

In 2016, the American Statistical Association (ASA) issued a statement on "Statistical Significance and p-values." Below is Chaps. 2 and 3 of the statement.

ASA Statement on Statistical Significance and p-Values

2. What is a p-value?

Informally, a p-value is the probability under a specified statistical model that a statistical summary of the data (e.g., the sample mean difference between two compared groups) would be equal to or more extreme than its observed value.

3. Principles

1. p-values can indicate how incompatible the data are with a specified statistical model.
2. p-values do not measure the probability that the studied hypothesis is true, or the probability that the data were produced by random chance alone.
3. Scientific conclusions and business or policy decisions should not be based only on whether a p-value passes a specific threshold.
4. Proper inference requires full reporting and transparency.
5. A p-value, or statistical significance, does not measure the size of an effect or the importance of a result.
6. By itself, a p-value does not provide a good measure of evidence regarding a model or hypothesis.

Chapter 2
The Invisible Hand of God

Significant yet Meaningless: Academic Journals Filled with Redundant Papers.

2.1 Is There an Effect in Dieting Methods?/Crash Course in Statistical Terms

In this section, we will first briefly review some basic concepts of statistics. Many of these concepts may already be familiar from middle and high school studies, but please recall them before proceeding to the main discussion.

> **Dieting Problem 1**: To investigate the effectiveness of a certain diet method, we had 20 women participate. The weights of participants before and after the program, as well as the difference in weight, are shown in Table 2.1. Now, is this diet method effective?

In statistics, objective measurement is emphasized. **Measurement** is assigning numerical values to **observational subjects** such as respondents or objects, based on predetermined operations. The numbers assigned through measurement are called **measured values**. The collection of these measured values is referred to as **data**. A collection of measured values that express one aspect of the observational subject's characteristics is called a **variable**. In Table 2.1, there are three variables: "Weight Before," "Weight After," and "Weight Difference."

H. Toyoda, *Statistical Significance and the PHC Curve*,
https://doi.org/10.1007/978-981-97-7748-8_2

Table 2.1 Differences in weight before and after participation in the diet (kg)

Participant number	1	2	3	4	5	6	7	8	9	10
Weight before	53.1	51.5	45.5	55.5	49.6	50.1	59.2	54.7	53.0	48.6
Weight after	48.3	45.2	46.6	56.6	41.2	44.6	51.9	55.5	45.4	47.6
Weight difference	4.8	6.3	− 1.1	− 1.1	8.4	5.5	7.3	− 0.8	7.6	1.0
Participant number	11	12	13	14	15	16	17	18	19	20
Weight before	55.3	52.6	51.7	48.6	56.4	42.9	50.3	42.4	51.2	39.1
Weight after	50.6	54.5	49.0	43.9	53.8	40.1	52.8	35.3	55.6	38.0
Weight difference	4.7	− 1.9	2.7	4.7	2.6	2.8	− 2.5	7.1	− 4.4	1.1

For the ith observational subject (female), the "Weight Before," "Weight After," and "Weight Difference" are denoted as x_{bi}, x_{ai}, and $x_{\Delta i}$ respectively. There is a relationship between these three variables as follows:

$$x_{\Delta i} = x_{bi} - x_{ai}, \quad (i = 1, \ldots, n) \tag{2.1}$$

For example, for $i = 1$, it is 4.8 kg = 53.1 kg − 48.3 kg. The small letters serve as subscripts; here i is a subscript distinguishing the observational subjects. When the observational subjects are people, i can also be referred to as the participant number.

The number of observational subjects is denoted as n. In this case, $n = 20$. The notation $(i = 1, \ldots, n)$ indicates that the subscript i varies from 1 to 20. The small letters "difference," b, a are subscripts to distinguish the variables.

Handling all three variables simultaneously can be complicated, so for the time being, we will focus on the "Weight Difference" $x_{\Delta i}$ to explain the methods of elementary statistical analysis. The analysis of data begins with a careful observation of each measurement. Observing the "Weight Difference" visually, we find that there are six negative values. It can be seen that 6 out of 20 people gained weight after the diet program.

Observing data visually is important. However, since there are limitations to visual observation, we next examine the **distribution of the variable**. As mentioned in Chap. 1, **distribution** refers to the pattern of how and where the data are observed.

2.1.1 *Histograms*

There are two types of data. Data that are discrete and countable, like the number of successes in an "Extrasensory Perception Experiment," are called **count data**. Data that measure continuous quantities, like the "Weight Difference," are called **measurement data**. To examine the distribution of measurement data, creating a histogram like in Fig. 2.1 is effective.

Fig. 2.1 Histogram of "Weight Difference" (kg)

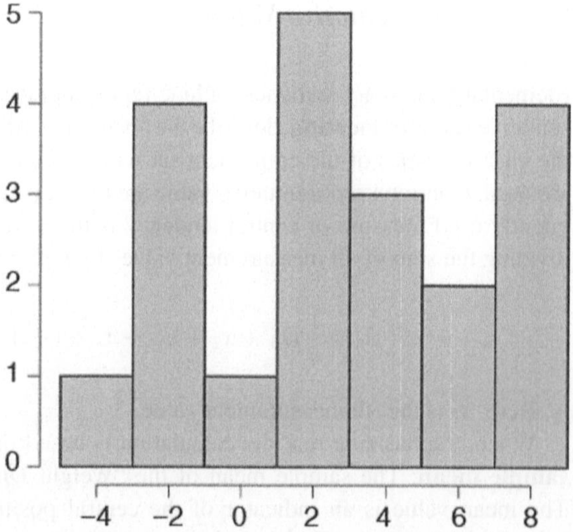

A **histogram** is a statistical graph with the class intervals on the horizontal axis and the frequency on the vertical axis. Here, **class interval** refers to the range of measurement values, and **frequency** refers to the number of measurements observed in each class. For example, the far-left bar indicates that there is one measurement value observed in the interval from − 5 kg to less than − 3 kg. The length of the interval is called **class width**. In Fig. 2.1, the class width is 2 kg. The middle value of a class is called the **class value**. For the far-left class (bar), the class value is − 4 kg.

Representing the characteristics of data in an easily understandable way through statistical graphs is called **graphical summary**.

2.1.2 *Numerical Summary*

A graphical summary is convenient as it allows us to understand the state of the data at a glance. However, it lacks simplicity. Therefore, numerical indicators are used to concisely describe the characteristics of the data. As we learned in Chap. 1, when viewing data as independent variables, the function defined in Eq. (1.3) is called a **statistical measure**. Among statistical measures, those used for test of significance are specifically called **test statistics**.

Here, we will learn about different statistical measures. These are measures used to condense the properties of data, known as **summary statistics**. Furthermore, summarizing the characteristics of data using summary statistics is called **numerical summary**. Numerical summary is a term that pairs with graphical summary.

2.1.3 Representative Value

Elementary summary statistics include representative value and dispersion. Representative value, or **location**, describe the position of a distribution. When representing the characteristics of the entire data set with a single number, representative value are used. Common representative value are mean, median, and mode. The most frequently used measure of central tendency is the mean. The **mean** is calculated by dividing the sum of all measurement values by n[1]:

$$\bar{x} = \frac{1}{n}(x_1 + x_2 + \cdots + x_i + \cdots + x_{n-1} + x_n) \tag{2.2}$$

Here, x_i is the ith measurement value.

When emphasizing that the calculation is based on data, $\bar{x}$ is referred to as the **sample mean**. The sample mean of the "Weight Difference" was $\bar{x}_\Delta = 2.74$ kg. The mean value is an indicator of the central position of the distribution. Also, $\bar{x}_b = 50.56$ kg, $\bar{x}_a = 47.83$ kg.

Arranging data in ascending order is called **sorting**. For example, the sorted data for "Weight Difference" is as follows:

```
- 4.4, - 2.5, - 1.9, - 1.1, - 1.1, - 0.8, 1.0, 1.1, 2.6, 2.7,
  2.8,   4.7,   4.7,   4.8,   5.5,   6.3, 7.1, 7.3, 7.6, 8.4
```

Consider an trial where a measurement value is drawn from this distribution. When the probability of observing a value less than or equal to x is $\ddot{a}$, then x has a **cumulative probability** of $\ddot{a}$. Alternatively, x is called the **$\ddot{a}$ x 100th percentile**. For example, since -4.4 is the smallest value, the probability of observing a value less than or equal to -4.4 is equal to the probability of observing -4.4. With $n = 20$, this probability is $1 \div 20 = 0.05$. Therefore, the cumulative probability of -4.4 is 0.05, and it is called the 5th percentile. Similarly, the lower cumulative probability for -2.5 is 0.1. It can be politely said as the lower 10th percentile.

The second measure of central tendency, the **median**, is the middle value of the sorted data. If n is odd, the median is the $(n + 1)/2$th measurement value. For example, if $n = 19$, the 10th measurement value in the sorted data is the median. If n is even, the median is the average of the $n/2$th and $(n/2) + 1$th measurement values. For the "Weight Difference" data, $n = 20$, so the median is the average of the 10th and 11th measurement values, 2.75 kg $(= (2.7 + 2.8)/2)$.

The third measure of central tendency mentioned, the **mode**, is the class value with the highest frequency. Observing Fig. 2.1, in the case of a class width of 2 kg, the frequency in the interval from 1.0 kg to less than 3.0 kg is the highest at 5. Therefore, the mode is 2.0 kg.

---

[1] $\bar{x}$ is read as "x-bar."

2.1.4 Dispersion

A summary statistic that describes how much the measurement values are spread out on average from the central location of the distribution is called **dispersion**. Common summary statistics for dispersion are variance and standard deviation. The statistic obtained by subtracting the mean $\bar{x}$ from each measurement value x_i, squaring the result, and then dividing the sum of these squared values by $n-1$

$$s^2 = \frac{1}{n-1}\left((x_1-\bar{x})^2 + \cdots + (x_i-\bar{x})^2 + \cdots + (x_n-\bar{x})^2\right) \qquad (2.3)$$

is called **variance**. When emphasizing that it is calculated from the data, it is called **unbiased variance**. The unbiased variance for "Weight Difference" is $s_\Delta^2 = 14.50$.

However, the unit of variance is the square of the measurement values. Therefore, as a measure of average dispersion, the value of variance cannot be concretely interpreted. To compensate for this drawback, the square root of the variance is taken as follows

$$s = \sqrt{s^2}$$

and converted back to the original unit of measurement.

This is called the **standard deviation**. It is sometimes abbreviated as sd. A large (small) standard deviation indicates that, on average, measurement values are observed far from (close to) the mean. The standard deviation for "Weight Difference" was $s_\Delta = 3.81$ kg. This indicates that, on average, the measurement values are spread out approximately 3.8 kg from the mean.

2.2 Test of the Mean Difference of Paired Two Groups

The test of significance is not limited to the test of population proportions. Here, to familiarize ourselves with the concept of tests, we introduce the test of the difference in population means, which, along with the test of population proportions, is one of the most frequently used tests.

Now, let's return to our initial question. Do the data from Table 2.1, showing an average weight loss of 2.74 kg among 20 people, truly indicate the effectiveness of this diet method? The answer is not necessarily. This time, the sample mean of weight loss happened to be 2.74 kg, but a repeat experiment might yield a negative value.

The most academically important thing is the reproducibility of a phenomenon. It is crucial that the average weight loss assumed under countless repeated experiments in the same conditions is significantly large.

The average mean mentioned above differs from the sample mean $\bar{x}$ calculated from the current data. To distinguish between them, let's call the average mean the **population mean** μ (mu). Academically, the interest lies in the value of the population mean μ, not in the sample mean $\bar{x}$.

What we are discussing here is the **Test of the Mean Difference of Paired Two Groups**. What does "mean difference of two groups" in the test name mean? It refers to the difference in population means of two groups: "Weight Before" and "Weight After." Moreover, "paired" implies that **"Weight Before" and "Weight After" are measured from the same individuals**.

Step 1 involves setting the null hypothesis H_0 and the alternative hypothesis H_1.

The null hypothesis is set as the population means of the two groups being equal,

$$H_0 : \mu_b = \mu_a \quad \text{or} \quad \mu_b - \mu_a = 0 \tag{2.4}$$

μ_b is the population mean of "Weight Before." μ_a is the population mean of "Weight After." However, it is known that there is a relationship between the two population means, $\mu_\Delta = \mu_b - \mu_a$. μ_Δ is the population mean of "Weight Difference." Utilizing this relationship, the null hypothesis can also be rewritten as

$$H_0 : \mu_\Delta = 0 \tag{2.5}$$

Equations (2.4) and (2.5) are equivalent in their expression. The hypothesis implies that the population mean of weight reduction is zero. In other words, the hypothesis suggests that the diet method has no effect at all. The alternative hypothesis, therefore, is $H_1 : \mu_\Delta \neq 0$.

Step 2 involves calculating the test statistic assuming H_0 is true.

In the "premonition of the future" trial, we used a binomial distribution for testing, and the sample proportion as the test statistic. Here, we will conduct a test using the normal distribution. The **normal distribution** is the most frequently used probability distribution to approximate the distribution of various continuous variables that occur around us. It is suitable for describing data that have a higher frequency around the mean and a decreasing frequency as it moves away from the center.

When a variable x follows a normal distribution, it is denoted as

$$\text{Normal distribution}(x|\mu, \sigma)$$

The shape of the normal distribution is determined by μ and σ. Here, μ is the **population mean**, and σ is the **population standard deviation**.[2] The vertical bar | in the parenthesis is read as "given." It means that x follows a normal distribution under the condition given by μ and σ.

Generally, numerical indicators that determine the shape of a probability distribution are called **parameters**. Examples of parameters are the population proportion π in the binomial distribution, and the population mean μ and population standard deviation σ in the normal distribution.

[2] σ^2 is called the **population variance**.

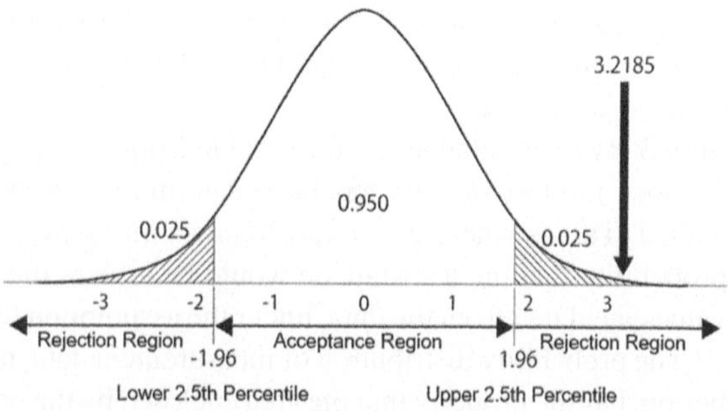

Fig. 2.2 Standard normal distribution

Since the shape of a normal distribution is determined by μ and σ, it is convenient to transform all normal distributions to have $\mu = 0$ and $\sigma = 1$ for consistent comparison. For this purpose, a linear transformation is applied to x

$$\frac{x - \mu}{\sigma} \tag{2.6}$$

Subtracting the mean from the variable x and dividing by the standard deviation. This transformation is called **standardization**. The standardized value is known to follow

$$\text{Normal distribution}(|\mu = 0, \sigma = 1) \tag{2.7}$$

The normal distribution with a population mean of $\mu = 0$ and a population standard deviation of $\sigma = 1$ is called the **standard normal distribution**. The standard normal distribution is shown in Fig. 2.2.

Let's assume that the "Weight Difference" from Eq. (2.1) follows a normal (not standardized) distribution

$$\text{Normal distribution}(x_{\Delta i}|\mu_\Delta, \sigma_\Delta) \tag{2.8}$$

where μ_Δ and σ_Δ are the population mean and standard deviation of $x_{\Delta i}$, respectively. Now, imagine that we repeat a trial with n observations countless times, assuming that the null hypothesis (2.5) is true. Further, imagine that for each experiment, a certain statistic[3]

$$z = \frac{\bar{x}_b - \bar{x}_a}{s_\Delta} \times \sqrt{n} = \frac{\bar{x}_\Delta}{s_\Delta} \times \sqrt{n} \tag{2.9}$$

is calculated and recorded each time. Here, s_Δ is the standard deviation of $x_{\Delta i}$, based on the square root of the unbiased variance. Let's call s_Δ the **standard deviation of the difference score**.

In reality, the experiment is conducted only once, so this is purely hypothetical. This hypothetical distribution of the statistic is called the **sampling distribution**. It is

[3] $\bar{x}_b - \bar{x}_a = \bar{x}_\Delta$.

known that the sampling distribution of z can be approximated by the standard normal distribution.[4,5] Therefore, z can be used as the test statistic for the null hypothesis (2.5).

Step 3 involves calculating the p-value from the sampling distribution.

Once the test statistic has been determined, its probabilistic evaluation is conducted. This is where the p-value comes into play. The ***p*-value** is defined as "the probability that the test statistic would be as extreme as, or more extreme than, the value calculated from the data, under the assumption that the null hypothesis is true."

The probability distribution of measurement data, not limited to the normal distribution, has the property that the area enclosed by the graph's curve and the horizontal axis equals one. When a specific interval is delineated, the area under the curve within that interval corresponds to the probability of observing values in that range.

Please refer to Fig. 2.2. In the standard normal distribution, 1.96 is called the upper 2.5th percentile. The probability of observing a value greater than or equal to 1.96 is 0.025. The cumulative probability from below is 0.975. -1.96 is referred to as the lower 2.5th percentile. The probability of observing a value less than or equal to -1.96 is also 0.025. Combined, they total 5%. Expressed using the absolute value, it becomes

$$p(1.96 \leq |z|) = 0.05$$

The probability of observing values corresponding to the hatched areas on the upper and lower sides of Fig. 2.2 is 5%. Thus, this region constitutes the rejection region for a test at the 5% level.

The test statistic for the diet experiment was calculated as

$$z = \frac{50.57 - 47.83}{3.81} \times \sqrt{20} = \frac{2.74}{3.81} \times \sqrt{20} = 3.2185 \tag{2.10}$$

As shown in Fig. 2.2, 3.2185 is a value within the rejection region. 3.2185 is much further to the right than the upper 2.5th percentile of 1.96. The probability that the absolute value of the test statistic z becomes greater than 3.2185 is the p-value. Therefore, $p < 0.05$.

Step 4 involves deciding whether to reject or accept the null hypothesis H_0.

In test of significance, if a p-value is small and an unlikely fact is observed, the conclusion is drawn that, rather than the null hypothesis being true and an unlikely

---

[4] More precisely, z follows a probability distribution known as the t-distribution with degrees of freedom $df = n - 1$. Here, the degrees of freedom df are a numerical indicator that determines the shape of the t-distribution and are parameters of the t-distribution. As it is clear from the definition, as n increases, df also increases. As the degrees of freedom df become large, the t-distribution approaches the standard normal distribution. For $n = 20$, a practical approximation can be obtained by the standard normal distribution.

[5] In most statistical education dealing with test of significance, students are not taught "why the test statistic follows ** distribution." They are instructed to "just remember it without thinking about the reason," which unfortunately puts a lid on the fresh intellectual curiosity of students and sends a strong message that "inferential statistics is just about memorization." The educational adverse effects of this are discussed in Chap. 7.

value being observed, it is more likely that the "null hypothesis is false." This is what we refer to as rejecting the null hypothesis. Since $p < 0.05$, we reject the null hypothesis as per Eq. (2.5) and conclude that the sample mean of "Weight Difference" $\bar{x}_\Delta = 2.74$ kg is statistically significant.

2.3 Statistically Significant but Scientifically Meaningless

The test statistic z, as can be seen from Eq. (2.9), is calculated based on three factors: standard deviation, sample mean, and n (number of observations). For the weight loss issue, considering only the upper side, if $1.96 < z$, it is statistically significant. Inserting this into Eq. (2.9) and transforming it, we get

$$\bar{x}_\Delta > \frac{1.96 \times s_\Delta}{\sqrt{n}} \tag{2.11}$$

If this equation holds, it is "statistically significant." The sample mean $\bar{x}_\Delta$ fundamentally reflects the effectiveness of the diet method.

However, the number of observations n (aside from budgetary constraints) can be freely determined by the researcher. Therefore (and this is very important), n is completely irrelevant to the scientific nature or usefulness of the diet method.

The sample standard deviation s does indicate some aspects of the diet method, but it does not directly reflect its effectiveness. For now, let's keep the sample standard deviation calculated from the data at $s_\Delta = 3.81$. This simplifies the equation without changing the essence of what we are discussing. The result is

$$\bar{x}_\Delta > \frac{1.96 \times 3.81}{\sqrt{n}} = f(n) \tag{2.12}$$

This equation shows how much the average weight loss needs to be, in kilograms, to be considered "significant." In this sense, it's an important formula that influences the publication of research.

This equation is a function of the number of observations n. Moving n around, we get

$$\begin{aligned}
\bar{x}_\Delta &> 1.670 = f(n = 20) \\
\bar{x}_\Delta &> 1.056 = f(n = 50) \\
\bar{x}_\Delta &> 0.747 = f(n = 100) \\
\bar{x}_\Delta &> 0.236 = f(n = 1000) \\
\bar{x}_\Delta &> 0.106 = f(n = 5000) \\
\bar{x}_\Delta &> 0.024 = f(n = 100{,}000)
\end{aligned}$$

The first equation is for $n = 20$, the actual experimental condition. It indicates that if the average weight loss is more than 1670 g, then this experiment is "statistically significant."

What happens if the number of subjects increases to 100? If the average weight loss is more than 747 g, the paper can claim, "This diet method is significant." With $n = 100{,}000$, a $\bar{x}_\Delta$ of just over 24 g is "statistically significant." However, a diet method resulting in just 24 g of weight loss is scientifically meaningless.

As the number of observations n increases, the required average weight loss for paper acceptance can approach 0 g indefinitely. The number of observations n, which the researcher can freely determine and is completely unrelated to the scientific properties of the diet method, essentially sets the acceptance criteria for the paper. The same situation has occurred as with Bem's test of proportions.

What about when n is small? For example, if $n = 5$, then

$$\bar{x}_\Delta > 3.34 = f(n = 5)$$

In this case, even if the average weight loss is 3.3 kg, it would not be significant.

2.4 The Null Hypothesis Is Scientifically False

The null and alternative hypotheses were respectively

$$\mu_\Delta = 0, \quad \mu_\Delta \neq 0$$

The null hypothesis is a point, a mathematical singleton, whereas the alternative hypothesis is almost the entire number line, excluding that one point.

It is mathematically permissible to assume "the null hypothesis is true," but scientifically, this is not feasible. In the language of analysis, a branch of mathematics, the null hypothesis is almost everywhere (a.e.) covered by the alternative hypothesis. To be covered mathematically a.e. means, scientifically, it is included. Even if $\mu_\Delta = 1$ pg (one trillionth of a gram), the alternative hypothesis is true, and the null hypothesis is false. If you practice flipping a coin intending to get heads, you can increase the frequency of heads by about 0.3%. Just exhaling reduces your weight. If the precision of measurements below the decimal point is increased, the null hypothesis is scientifically false. Even before collecting data, the null hypothesis is strictly false.

The property that "even if it is academically meaningless, the alternative hypothesis is true, and as the number of observations n increases, it will inevitably be rejected" is a common characteristic of most significance tests used in peer-review judgments. This is an absurd situation.

To overcome such drawbacks, it is necessary to link statistical indicators with academic value. The essentially important matter is to confirm that the weight has decreased sufficiently enough to be practically meaningful as a diet method. In other words, a methodology is needed that can actively confirm that the value of the population mean μ_Δ is sufficiently large.

2.5 The p-Value Question in the National Medical Examination

Teacher So far, we've covered the basics of tests. Do you understand everything?

Student It's quite difficult, but I've read through it several times, so I think I understand it up to this point.

Teacher Alright, let's test your understanding with a question. Please select the correct answer for the question in Fig. 2.3.

Student This question was actually asked in the National Medical Examination in Japan for physicians, wasn't it? A test of significance question in a medical exam.

Teacher Yes. Statistics are applied in various fields of study, so even physicians need knowledge of statistics. Pretend you're a medical student and give it a try.

Student Okay! I'll do my best. Hmmmm ... <A while later> I don't know. I can't solve this problem.

Teacher No, that's quite alright. The correct answer is announced as option e, but it's not the right answer. There is no correct answer to this question, so it can't be solved.

Student When I read option e, I thought, "The probability of falsely concluding there is a difference when the null hypothesis is true is the significance level", which isn't 3.6% p-value.

Teacher Exactly.

Student So there are mistakes in the National Medical Examination too.

Teacher Exams are made by humans, so mistakes are somewhat inevitable. What we need to focus on here is the possibility that **this was not a mistake.**

In a publication by a leading figure in the Statistical Society [7], the incident of this erroneous question is described as follows:

"I explained to several prominent physicians in the medical community and at Kurume University School of Medicine that it was a mistaken question, but they did not understand and I was shocked. Understanding that answer e is not correct requires basic statistical knowledge, which they lacked."

Student Perhaps it wasn't a mistake in the question. Are there doctors out there writing papers without understanding the meaning of p-values? That's kind of scary.

Teacher It varies from person to person. But it's unreasonable to assume 'Null hypothesis H_0: p-values are an easy-to-understand concept' is true, yet prominent medical professionals can't understand it.

Student When an unlikely fact is observed, we judge that it's not that the null hypothesis is true and an unlikely value was observed, but rather that "the null hypothesis is false." So we reject the null hypothesis and accept that "p-values are a difficult concept to understand," right? (laughs).

43 To investigate the therapeutic effects of newly released antibiotic A on pneumonia, newly admitted pneumonia patients were divided into two groups: a group treated with antibiotic A (Group A) and a group treated with the existing antibiotic B (Group B). The therapeutic effects were compared in terms of the duration of hospitalization. The results are shown in the table.

	Group A	Group B	p value
Number of subjects	198	201	
Length of hospitalization (mean)	8.1days	9.6days	0.036

Which of the following interpretations of the results is correct?
a. Group A has an average hospitalization duration that is 3.6% shorter than Group B.
b. The margin of error for the average hospitalization duration in Group A is within 3.6%.
c. The probability that Group A has a shorter hospitalization duration than Group B is 3.6%.
d. 96.4% of patients in Group A had a shorter hospitalization duration than the average for Group B.
e. The probability of incorrectly concluding that there is a difference in hospitalization duration between Group A and Group B when there is none is 3.6%.

Fig. 2.3 The National Medical Examination in Japan 105th session (2011) problem B, question 43

■ Typical example of misunderstanding of *p*-value

1. The probability that the null hypothesis is true is the *p*-value. The probability that the hypothesis of no difference in population means is correct is the *p*-value.
2. The probability that the alternative hypothesis is true is 1 – *p*-value.
3. The probability of observing this experimental data when the null hypothesis is true is the *p*-value.
4. The probability of mistakenly rejecting H_0 when the null hypothesis is true is the *p*-value.
5. The probability of obtaining a sample mean difference equal to or more extreme than the observed value is the *p*-value.
6. I forgot the meaning. Anyway, it's just a happy probability if it's less than 5%.

Fig. 2.4 Typical misunderstandings of *p*-value

2.6 The *p*-Value Is Probabilistic but Abstract

Teacher I have been teaching test of significance to psychology students for about 30 years. I have almost given up hope of getting them to understand *p*-values.

Student Maybe today's students aren't serious enough, and they just don't get it …

Teacher No, the students were excellent. The problem lies in the abstract nature of the *p*-value concept, which is hard to grasp. Even if they understand it right after being taught, many students misunderstand it a few years later. Typical misunderstandings of the *p*-value are summarized in Fig. 2.4.

Teacher It's difficult to explain that the first and second points represent different types of misunderstandings. Even the misunderstandings are misunderstood.

Student The fourth one is the misunderstanding in the earlier Japanese National Medical Examination question. But professor, the fifth point seems correct as an explanation of the p-value. There was a sentence with the same meaning in the "ASA Statement on Statistical Significance and p-values."[6]
Teacher It's true that there is a sentence[7] with the same meaning, but it is incorrect as an explanation of the p-value.
Student Why is that? Are you saying the ASA's explanation is wrong?
Teacher Look at the test statistic formula (2.9). $\bar{x}_\Delta$ is the sample mean of the difference between "before weight" and "after weight." It was 2.74 kg.
Earlier, for simplicity, we kept s_Δ at 3.81 kg. However, in reality, as we imagine repeating the experiments many times, s_Δ also changes. It's possible to imagine a sampling distribution for s_Δ too.
Let's say one set of these imagined retests had $\bar{x}_\Delta = 2.8$ kg, $s_\Delta = 4.0$ kg. The test statistic would be

$$z = (2.8/4.0) \times \sqrt{20} = 3.130495 < 3.2185 \tag{2.13}$$

which is smaller than the value in formula (2.10).
Even if the sample mean difference in the imagined retest is greater than the experimental data (2.74 kg < 2.8 kg), if s_Δ is also larger (3.81 kg < 4.0 kg), the z value could be smaller than the value actually calculated from the experimental data. That's why the p-value is not "the probability of obtaining a sample mean difference equal to or more extreme than the observed value."
Student So, even the statistical experts at ASA can't explain the p-value well. No wonder doctors, let alone students, misunderstand it.
By the way, teacher! I just thought of something. The mistaken answer options a, c, and d in Fig. 2.3 seem very important, don't they? They're more useful medically (essentially scientifically) than just whether it's significant or not.
Teacher You've grasped the essence. Questions like "By what percentage does a new treatment method shorten the average hospital stay?" "What percentage of patients have a shorter hospital stay with the new treatment compared to the conventional method?" "Where does the average stay of the conventional method fall in the distribution of stays with the new treatment?" and "How reliable are these estimates?" are essentially important medically.
They are clearly more useful for doctors than the information provided by test of significance. Options a, c, and d unintentionally represent the information that medicine truly needs.[8] These options are the unconscious wishes of the doctors who set the question, yearning for statistical understanding. Yes, I'm sure of it.

---

[6] See column in Chap. 1.

[7] The original text is as follows: Informally, a p-value is the probability under a specified statistical model that a statistical summary of the data (e.g., the sample mean difference between two compared groups) would be equal to or more extreme than its observed value. The translation by the Japan Biometric Society is accurate.

[8] Option a asks for the probability of a hypothesis regarding the **distribution of ratios**, option c for the probability of a hypothesis regarding the **rate above a threshold**, option d for the probability

2.7 The Invisible Hand of God

There is a discrepancy between statistical significance demonstrated by testing and a difference that is academically meaningful. The null hypothesis for the binomial test of proportions was

$$\pi = 0.5 \tag{2.14}$$

For the test of the mean difference of paired two groups, it was

$$\mu_b = \mu_a \tag{2.15}$$

Rejecting these hypotheses has led to the publication of articles in academic journals, showing statistical significance.

However, in substantive science, a 53% "predictive ability" is negligible, and for a new diet method, showing a significant difference in weight loss of just 24 g over several months is meaningless. The research hypothesis that substantive science seeks, for example in the two cases mentioned earlier, is

$$\pi - 0.5 > c \quad \text{or} \quad \mu_b - \mu_a > c \tag{2.16}$$

The reality-useful difference is when the left side of Eq. (2.16) is greater than a (statistically undetermined) threshold c.

If the left side of Eq. (2.16) is greater than 0 but less than c, it is substantively meaningless, but from the viewpoint of significance testing, it is a state where the alternative hypothesis is true (the null hypothesis is false).

This is the area where the procedure of significance testing diverges from the original research objectives. Should we then avoid this area through the analyst's conscience and good sense? No, it is practically impossible, which is why criticism of the p-value is recurrent.

If academic articles continue to be published based on significant results from significance testing, this discrepancy area becomes increasingly populated with articles, eventually leading to a rise in the proportion of statistically significant but meaningless papers. This is an inherent property of the p-value. Why is this so? It is because the invisible hand of God, as discussed by Adam Smith in "The Wealth of Nations," comes into play.

2.7.1 Focusing Only on Profit Leads to Appropriate Pricing

In a free-market economy, sellers can freely determine the quality and price of their products. They are allowed to prioritize their own profit when setting the quality and

---

of a hypothesis regarding the **non-overlap degree**. These can be analyzed in practice. We will deal with this in Chap. 6. Readers interested in more details may refer to the literature [8].

price. Everyone wants to sell at a high price, but if the price is too high, it won't sell, resulting in a loss. If priced too low, although it may sell, the profit margin becomes too small.

If a high-quality product is offered, the probability of selling increases, but higher quality costs more to acquire, potentially reducing profit even if it sells. Adjusting the quality and price of the product, through countless trials of "sold," "not sold," "sold," "not sold," ... eventually leads the quality and price of the product to settle at an economically appropriate level. Adam Smith referred to this mechanism as the **invisible hand of God**.

2.7.2 *Focusing Only on Statistical Significance Leads to a High and Stable Proportion of Meaningless Papers in Journals*

What happens when academic articles continue to be published based solely on the significance of the results of significance testing? Here, "selling" is an analogy for publishing articles due to significant differences. "Good quality" of the product is analogous to the fulfillment of Eq. (2.16). "Cheapness of price" corresponds to the size of n, but note that the direction is the opposite.

Researchers can decide on n. With fewer data, significant differences are harder to obtain. If an article isn't published, the research is wasted. With massive data collection, even meaningless differences can become significant.

If a researcher can conduct a study that has academic value (fulfills Eq. (2.16)), significant differences can be obtained even with a small amount of data, enabling the publication of a paper. However, conducting a valuable study is a kind of creative feat. The cost involved is incomparably higher than the effort of data collection. Academic creation is clearly more costly than data collection.

In this situation, the community of researchers experiences countless trials of whether their studies show significant differences (whether they can be published or not). The invisible hand of God operates in this process. Over time, even meaningless papers start to show significant differences as n gradually increases.

Since data collection also incurs some costs, n won't increase indefinitely. Eventually, n will stabilize in a region where significant differences are empirically likely to be obtained. This n becomes the conventional and average number of observations in that research field. This mechanism is illustrated in Fig. 2.5.

As evidence, for instance, the literature [9] shows a trend that in research fields where the average difference in means is large, n tends to be smaller on average, and in fields where the average difference in means is small, n tends to be larger.

Every researcher has a heartfelt desire to publish their papers in academic journals. Since showing significant differences is a legitimate action within the rules, researchers and their papers naturally flock to this discrepancy without any malice. This is why focusing solely on statistical significance results in academic journals being filled with scientifically meaningless papers.

■ " The invisible hand of God "

1. Cheaper prices make it easier to sell (a larger n makes it easier to achieve significance). Higher prices make it harder to sell (a smaller n makes it harder to achieve significance).
2. High-quality products are easier to sell (findings that fulfill equation (2.16) are academically valuable and, of course, more likely to be significant).
3. However, high-quality products are scarce and hard to source (valuable discoveries, being creative feats, are extremely costly and difficult to achieve).
4. The seller decides the price (the researcher decides the number of observations n).
5. Merchants (researchers) accumulate countless experiences of selling or not selling (being published or not) in their struggle for survival.
6. The invisible hand of God operates.
7. The price settles at a level where low-quality goods can be sold (academically meaningless, but the average n in that field of research stabilizes at a level where significant differences can be obtained).

Fig. 2.5 Operation of the invisible hand of God when high-quality products are scarce and hard to source

2.8 Conclusion

Journals that automatically consider articles significant if the p-value falls below 5% are filled with academically meaningless papers due to the invisible hand of God. Like pricing, in a life-or-death situation, the invisible hand acts quickly and surely. This is not a problem of researchers' ethics or conscience, but a natural consequence of the contradictions in the peer-review system.

The p-value is probabilistic yet abstract, making it difficult to interpret in a scholarly context. This leads to a tendency to mechanically handle the results as "significant." This also unconsciously accelerates the workings of the invisible hand. The abstract nature of the probability leads to misunderstandings that one has produced a high-quality result. To eliminate this drawback, it is crucial to use standards that researchers and reviewers in the field, who are not experts in statistics, can relate to and feel when making peer-review decisions.

For example, consider criteria such as how much a new treatment method reduces the average hospitalization period, what percentage of patients have a shorter hospital stay compared to the traditional method, and where in the distribution of hospitalization days of the conventional method does the new treatment method lie. Reviewing papers based on such criteria would prevent meaningless papers from being published in academic journals. Indicators that directly express the results of research and are relatable to users (doctors) are needed. Specific methodologies for this will be introduced in the latter part of this book.

Unfortunately, due to the invisible hand, the ratio ($a : b$ = athletes: humans) seems destined to have a higher proportion of humans. Now, have all the mysteries been solved? No. As astute readers may have noticed, this raises an even more puzzling mystery.

A significant difference is only a necessary condition for usefulness. However, statistically significant papers, even if scientifically meaningless, should be reproducibly significant in their meaninglessness. Yet, 75% of these papers are not even reproducible. Why is this so? This trick cannot be fully explained by the invisible hand alone. Are these 75% of papers an unknown creature, neither elite athletes nor mere humans? The mystery is still far from fully revealed.

Chapter 3
Tiger at the Front Gate, Wolf at the Back Gate

n must be predetermined with evidence provided (The phrase "Tiger at the front gate, wolf at the back gate" is a Japanese and Chinese idiom. It is an analogy for a situation where avoiding one danger leads to encountering another. It carries the same meaning as the English idiom "Between the hammer and the anvil," which similarly describes a scenario of being caught between two hazards.).

3.1 Efficacy of Antibacterial Drug A

Hospital Stay Issue: To investigate the efficacy of the newly developed antibacterial drug A, a comparative experiment with the existing antibacterial drug B was conducted. Newly admitted patients were randomly assigned to either the group receiving antibacterial drug A (Group A) or the group receiving antibacterial drug B (Group B), and the therapeutic effects were compared based on the duration of hospital stay. The results obtained are shown in Table 3.1.

In experiments, the intervention being tested is referred to as the **treatment**. In this experiment, the treatment is the administration of Antibiotic A. The group that receives the treatment is called the **Experimental Group**. However, it is difficult to confirm the effect of the treatment with only the experimental group. Therefore, to compare and contrast the effect of the treatment, Group B is set up as the **Control Group**.

However, if the experimental group consists only of robust patients, and the control group consists of frail patients, the comparison of hospitalization duration becomes meaningless. For this reason, the experimental group and the control group are made

H. Toyoda, *Statistical Significance and the PHC Curve*,
https://doi.org/10.1007/978-981-97-7748-8_3

Table 3.1 Summary statistics of hospitalization duration for patients treated with Antibiotics A and B

	Number	Average (days)	Standard deviation (days)
Experimental group (Group A)	24	8.1	2.3
Control group (Group B)	26	9.6	2.3

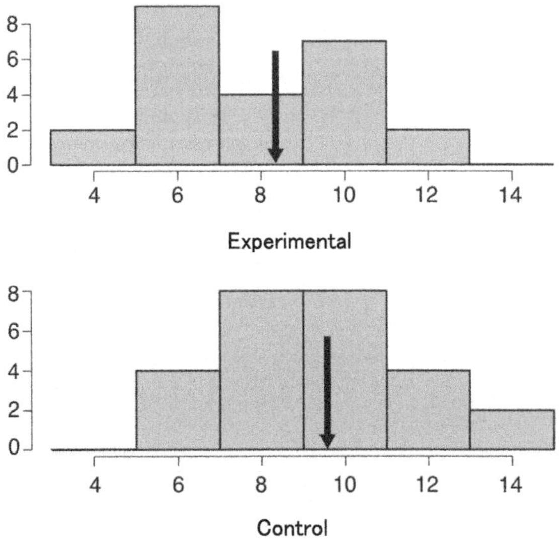

Fig. 3.1 Histogram of hospitalization duration for experimental and control groups

as similar as possible in conditions other than the treatment. This is called a **Controlled Experiment**. One strategy to achieve a controlled experiment is to randomly assign patients to the two groups. This is referred to as **Random Assignment**.

Figure 3.1 presents the histogram of hospitalization duration for both the experimental and control groups. The experimental group generally shows a shorter duration of hospitalization (distributed more to the left) compared to the control group. The positions of the arrows indicate the average hospitalization durations, which are 8.1 days and 9.6 days, respectively. The standard deviation of the hospitalization duration coincidentally matched up to the first decimal place, being 2.3 days.

From the average values and the shape of the histogram, can we conclude that administering the newly developed Antibiotic A results in a shorter average duration of hospitalization? Not necessarily; this does not always indicate effectiveness. In this instance, the average hospitalization duration happened to be 1.5 days shorter, but in a repeated experiment, it might turn out longer. What's important is the difference in the population mean of hospitalization duration when we assume an infinite number of repeated experiments under the same conditions. Let's perform a significance test.

Along with the test for the mean difference between two related groups learned in the previous chapter, another frequently used significance test is the **Test for Mean Difference Between Two Independent Groups**. We will use the latter here. In an experimental design for two related groups, we measured the weight of the same subject population before and after a diet, referring to them as the before group and the after group, respectively. In contrast, in Table 3.1, different patients are assigned to the experimental and control groups. Here, "independent" means that the measurements collected from the two different groups are unrelated to each other between the groups.

3.2 Test for Mean Difference Between Two Independent Groups

Step 1 involves setting up the null hypothesis H_0 and the alternative hypothesis H_1.

The null hypothesis is set as the population means of the two groups being equal,

$$H_0 : \mu_{\text{control}} = \mu_{\text{experimental}} \quad \text{or} \quad \mu_{\text{control}} - \mu_{\text{experimental}} = 0 \tag{3.1}$$

μ_{control} represents the population mean of the control group, and $\mu_{\text{experimental}}$ represents the population mean of the experimental group. The hypothesis implies that the population means of the hospitalization periods for Antibiotics A and B are exactly the same. The alternative hypothesis, being the negation, is $H_1 : \mu_{\text{control}} \neq \mu_{\text{experimental}}$.

Step 2 involves calculating the test statistic assuming H_0 is true.

Assuming that data from both groups come from normally distributed populations with a common standard deviation, let's say the experimental group has n_1 data points, and the control group has n_2 data points. For the antibiotic example, $n_1 = 24$, $n_2 = 26$.

Now, imagine repeatedly conducting experiments with the observed subject numbers n_1, n_2 as in the null hypothesis (Eq. 3.1), and in each of these experiments, calculate and record the following statistic:

$$z = \frac{\bar{x}_{\text{control}} - \bar{x}_{\text{experimental}}}{s_{\text{within}}} \times \sqrt{n^*}, \quad \text{where} \tag{3.2}$$

$$s_{\text{within}} = \sqrt{\frac{(n_1 - 1)s_1^2 + (n_2 - 1)s_2^2}{n_1 + n_2 - 2}} \tag{3.3}$$

$$n^* = \frac{n_1 n_2}{n_1 + n_2} \tag{3.4}$$

Remember, this is only an imagination, as in reality, the experiment is conducted only once. Here, s_{within} is a measure of the average spread within the groups. It is calculated by averaging the unbiased variances of both groups weighted according to the number of observations, and then taking the square root, giving us a common standard deviation for both groups. This is referred to as the within-group standard deviation. Since the standard deviations for both groups coincidentally were 2.3, we have $s_{\text{within}} = 2.3$. n^* is a function of n_1, n_2: the numerator in the square root is multiplication, while the denominator is addition, so as n_1 or n_2 increase, n^* also increases.

The imagined distribution of this statistic is called the **sampling distribution** of z. It is known that the sampling distribution of z can be approximated by a standard normal distribution.[1,2] Therefore, z can be used as a test statistic for the null hypothesis (Eq. 3.1).

Step 3 involves calculating the p-value from the sampling distribution.

Once the test statistic is determined, its probabilistic evaluation is performed. This is where the p-value is calculated. The ***p*-value** is defined as 'the probability that, under the assumption that the null hypothesis is true, the test statistic will take on a value as extreme or more extreme than the value calculated from the data.'

For the antibiotic experiment, the test statistic was found to be

$$z = 2.16 \tag{3.5}$$

In a standard normal distribution, $p(1.96 < |z|) = 0.05$ so it is significant at the 5% level ($p < 0.05$). 2.16 is within the rejection region. The probability of observing a value as extreme or more extreme is, from the standard normal distribution table,

$$p(2.16 < |z|) = 0.031$$

Therefore, the p-value[3] is 0.031.

Step 4 involves either rejecting or accepting the null hypothesis H_0.

In significance testing, if a p-value is observed to be less than 0.05, indicating an unlikely event under the assumption that the null hypothesis is true, it is concluded that the null hypothesis is likely false. Since $p < 0.05$, we reject the null hypothesis (Eq. 3.1) and conclude that the difference in sample means of hospitalization duration of 1.5 days (9.6 − 8.1) is statistically significant.

---

[1] More precisely, z follows a probability distribution known as the t-distribution with degrees of freedom $df = n_1 + n_2 - 2$. As the degrees of freedom df increase, the t-distribution can be approximated by the standard normal distribution. In the case of the antibiotic example with $df = 48(24 + 26 - 2)$, an adequate approximation is achieved with the standard normal distribution.

[2] In significance testing, other test statistics that follow the chi-square distribution or F-distribution are also used. However, in the education of significance testing, often the reasons why test statistics like those in Eq. (3.2) follow certain distributions are not taught. This leads students to adopt a learning attitude that "test statistics are something to be memorized," contributing to a sense of learned helplessness about being unable to devise or improve analytical methods. This is very regrettable. We will discuss this issue in more detail in Chap. 7.

[3] If calculated precisely using the t-distribution with 48 degrees of freedom, without approximating with the normal distribution, the p-value would be 0.036. This shows that the normal approximation is very accurate.

3.2.1 Difference in Average Hospitalization Days Required for Statistical Significance

Let's consider how much reduction in the average hospitalization days is needed for the difference to be statistically significant. Since we are interested in reducing hospitalization time, we only consider the upper side, so if $1.96 < z$, it becomes statistically significant. s_{within} reflects a part of the characteristics of hospitalization duration but does not directly reflect the efficacy of the new drug. Therefore, we will use the calculated $s_{\text{within}} = 2.3$ for now. To simplify the situation, let's assume that the number of observed subjects in both the experimental and control groups is the same, $n(= n_1 = n_2)$. Substituting this into Eq. (3.2) and transforming the equation, we get

$$\bar{x}_{\text{control}} - \bar{x}_{\text{experimental}} > \frac{1.96 \times 2.3}{\sqrt{n^*}} = f(n) \tag{3.6}$$

This becomes a function of n. If this equation holds, it is "statistically significant." The left side of the equation represents the average reduction in hospitalization days, essentially reflecting the efficacy of the antibiotic. In contrast, the number of observed subjects is determined by the researcher and is entirely unrelated to the usefulness of the antibiotic. Generally, the number of observed subjects is unrelated to the effect of the experimental treatment.

Equation (3.6) shows how much reduction in the average hospitalization duration is needed to declare the results "significant" in a paper. This is a critical equation in academic journals where significance is emphasized, determining whether research is published. It is indeed strange that this equation, reflecting the efficacy of an antibiotic, is expressed as a function of the unrelated variable n.

Let's see what happens when we vary n:

$$\bar{x}_{\text{control}} - \bar{x}_{\text{experimental}} > 4.8\,\text{h} = 0.20\,\text{days} = f(n = 1000)$$
$$\bar{x}_{\text{control}} - \bar{x}_{\text{experimental}} > 92\,\text{min} = 0.064\,\text{days} = f(n = 10{,}000)$$
$$\bar{x}_{\text{control}} - \bar{x}_{\text{experimental}} > 29\,\text{min} = 0.020\,\text{days} = f(n = 100{,}000)$$

For $n = 1000$, an average reduction of 4.8 h in hospitalization time means the experiment can be reported as "statistically significant" in a paper.

In the modern era of advanced networks and handling big data, dealing with data of $n = 100{,}000$ subjects is not uncommon. In such cases, an average reduction in hospitalization time of just 29 min is determined to be "statistically significant." Mathematically, the null hypothesis may be false, but medically, this is completely meaningless. However, it can be reported that "the difference in average values between the experimental and control groups is significant." This is the same issue that occurred with Bem's data in the test of proportions. This kind of drawback, which increases with n, is a common flaw in most significance tests.

Conversely, what if n is small? For example, if $n = 5$, then

$$\bar{x}_{\text{control}} - \bar{x}_{\text{experimental}} > 2.85\,\text{days} = f(n = 5)$$

In this case, even if the average reduction in hospitalization time is 2.5 days, it would not be statistically significant. If patients could be discharged 2 and a half days earlier, they would undoubtedly be very happy. They could watch movies, clean, and do various other activities. But this difference is not statistically significant.

3.3 Two Types of Errors Associated with Decision-Making

In significance testing, we assume the null hypothesis is true and reject it if something highly unlikely occurs. This unlikely probability is called the **significance level**. Often, a 5% significance level is used. However, stating that something occurs "only 5% of the time" also means that "it occurs in 5% of cases." Therefore, there is a chance that the null hypothesis is actually true, but is mistakenly rejected. Here, let's represent the significance level not as 5%, but generally as α. The significance level α is also known as the **risk level**. A test at the α level implies the risk of making an error in rejecting the null hypothesis even though it is true, with a probability of α. This error is called the **Type I error**. α represents the probability of committing a Type I error. The results of a test are not always correct.

3.3.1 Reasons Not to Further Reduce the Risk Level

If errors are inevitable, it seems better to have a smaller probability of them occurring. Is a test at the 1% level always better than a test at the 5% level? It would be problematic if incorrect papers are published. Should we, therefore, reduce the risk level α to something much smaller, like 0.1% or 0.001%?

That's not the case. If we reduce the probability α of committing a Type I error (rejecting the null hypothesis when it is actually true), we inadvertently increase the probability of committing a Type II error—failing to reject the null hypothesis when the alternative hypothesis is true. This is called the **Type II error**. The probability of committing a Type II error is denoted as β. This is a complex issue, so please refer to Fig. 3.2 for a clearer understanding.

With the null hypothesis, there is always the possibility of committing an error, whether it is accepted or rejected. Moreover, α and β have conflicting properties. With a fixed number of observed subjects, it is not possible to simultaneously reduce both. Reducing one will increase the other.

Why does reducing one increase the other? Let's explain this characteristic using the analogy of a court's decision on guilt or innocence. For example, let's set the

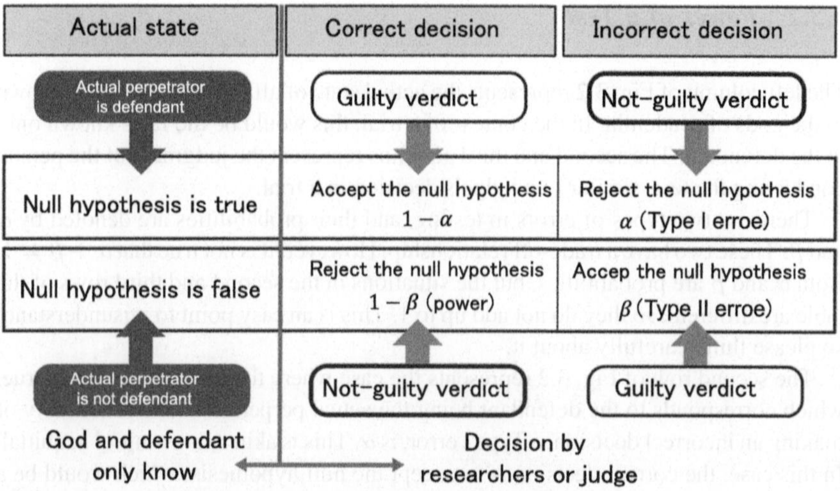

Fig. 3.2 Two types of errors associated with decision-making

null hypothesis as "Person A is the real perpetrator." In this case, accepting the null hypothesis equates to a "guilty" verdict, while rejecting it corresponds to a "not guilty" verdict.

What is the error that occurs with a probability of α? A Type I error is made when the null hypothesis is true, but it is incorrectly rejected. This error is like judging a guilty person as innocent—allowing the criminal to escape.

What about the error that occurs with a probability of β? A Type II error is made when the null hypothesis is false, but it is incorrectly accepted. This error is akin to judging an innocent person as guilty—a miscarriage of justice. Both α and β represent probabilities of errors we want to avoid.

Judges examine the evidence to assess the possibility of "Person A being innocent." A large test statistic z suggests that under the null hypothesis "Person A is the real perpetrator," there are many facts found that are unlikely to occur.

Imagine a judge who is a human rights advocate. If the z value is high, they will naturally discard the null hypothesis and deliver a not guilty verdict. Even if the z value is low, they tend to give a not guilty verdict to avoid wrongful convictions. In this case, the probability of letting the real criminal escape, α, is high, but the probability of wrongful conviction, β, is low.

Another judge is reluctant to deliver not guilty verdicts. When the z value is low, a guilty verdict is, of course, the outcome. Only when the z value is very, very high does this judge finally issue a not guilty verdict. In this case, the probability of letting the real criminal escape, α, is low, but the probability of wrongful conviction, β, is high. Thus, it is challenging to simultaneously minimize both α and β.

3.3.2 *Power of a Test*

The left column of Fig. 3.2 represents the actual state of affairs, which is only known to the gods of academia. In the context of a trial, this would be the facts known only to the defendant. The second and third columns represent the judgment of the person conducting the test, similar to a judge's decision in a trial.

There are two types of errors in testing, and their probabilities are denoted by α and β. These two have a trade-off relationship. However, it is not true that $\alpha + \beta = 1$. Both α and β are probabilities, but the situations in the second and third rows of the table are different, so they do not add up to 1. This is an easy point to misunderstand, so please think carefully about it.

The second row of Fig. 3.2 represents the case where the null hypothesis is true, which corresponds to the defendant being the actual perpetrator. The probability of making an incorrect decision, a Type I error, is α. This is akin to a wrongful acquittal. In this case, the correct decision is to accept the null hypothesis, which would be a guilty verdict. The probability of making this correct decision is not β, but $1 - \alpha$.

The third row of Fig. 3.2 represents the case where the null hypothesis is false, which corresponds to the defendant being innocent. The probability of making an incorrect decision, a Type II error, is β. This would be equivalent to a wrongful conviction. Therefore, the probability of making the correct decision is $1 - \beta$. The term $1 - \beta$ is known as **power of a test** or **statistical power**. It represents the probability of correctly rejecting the null hypothesis when it is false.

Since the null hypothesis $\mu_{\text{experimental}} = \mu_{\text{control}}$ represents a specific situation, α can be explicitly set, like at 5%. However, when the null hypothesis is false, various situations can arise, making it impossible to explicitly determine β. Consequently, the power of the test $1 - \beta$ is also unknown.

3.4 The Null Hypothesis Cannot Be Adopted

In the previous chapter, we discussed that statistical significance does not necessarily imply scientifically meaningful differences. Here, we will explain how to interpret results when they are not significant.

When calculating the z value under the assumption that the null hypothesis is correct, if the result is not unusual, the null hypothesis might be accepted. However, accepting the null hypothesis does not mean interpreting that the population means of the two groups are equal. The acceptance of the null hypothesis in statistical testing is different from the everyday use of the word "acceptance." This is a major point where significance testing can be easily misused.

In the article "Retire Statistical Significance" (2019) [2] published in the journal Nature, which was mentioned in the introduction and endorsed by over 800 scientists,

> We should never conclude there is "no difference" or "no association" just because a p value is larger than a threshold such as 0.05 or, equivalently, because a confidence interval includes zero.

This mistake is very common and serious. According to the same document, an examination of 791 papers published in five leading journals found that 51% of them had made this incorrect interpretation. Even professional researchers publishing papers in prestigious journals misused this concept more than half of the time.

Why is there such prevalent misuse? Are researchers incompetent? No, that is not the case. There is a genuine scientific need to substantiate "no difference" or "no effect." For example, if it can be shown that there is no difference in the condition of the experimental and control groups before administering Antibiotics A and B, it positively demonstrates that the controlled experiment has been conducted appropriately. This leads to the inadvertent misuse. To meet this urgent need, one must first define scientifically (not statistically) what constitutes a "no difference" or "no effect" state, and then confirm it statistically. Specific solutions to this issue are explained in Chap. 5 and onward.

3.4.1 Significance Means Finding an Alibi

Actual trials do not precisely represent significance testing.[4] The earlier example of a trial was intentionally set to differ from real trials. A large test statistic z was likened only to the situation where many facts unlikely under the null hypothesis "Person A is the real perpetrator" were found. The judge was only assessing the possibility of "Person A being innocent." In other words, significance testing does not collect or scrutinize evidence of guilt. Therefore, a guilty verdict cannot be rendered. In such cases, the verdict is suspended due to insufficient evidence of innocence. This is where it differs from a real trial.

The null hypothesis was "Person A is the real perpetrator." What would the z value, or test statistic, be in this case? It would be a fact that is unlikely to occur if we assume the null hypothesis is true, like finding an alibi.

An alibi corresponds to a large absolute value of the test statistic. Suppose the crime occurred in Tokyo, and an alibi was found that Person A was in Osaka at the same time. Rather than assuming Person A is the real perpetrator in Tokyo and was simultaneously in Osaka, the null hypothesis is rejected, and a verdict of innocence is rendered.

[4] More accurately, actual trials do not represent significance testing that deals with continuous population parameters like π or μ.

Thus, to **reject the null hypothesis** means to actually discard the null hypothesis, prepared for the risk of a Type I error. This usage is the same as in everyday language. A verdict of innocence is rendered, prepared for the possibility of letting the criminal escape. Antibiotic A is deemed better than the conventional product.

3.4.2 Attitude Reservation if No Alibi Is Found

But what about acceptance? Suppose under the assumption that the null hypothesis is true, nothing unusual happens. For example, what if no alibi is found? This is the situation where the absolute value of z is small, or $p > 0.05$.

Under the null hypothesis that Person A is the real perpetrator, not finding an alibi is neither unusual nor noteworthy. Does this mean we can conclude Person A is the real perpetrator? No, while an alibi could justify a verdict of innocence, the lack of an alibi alone does not warrant a verdict of guilt. It is merely a reservation of attitude.

To **accept the null hypothesis** does not mean to actively adopt it and utilize its findings. In significance testing, just because it is not significant does not mean we can assert there is no difference. Let alone claim the population means are the same. Discussions should not proceed on the premise that there is no difference. Instead, interpret or describe that no evidence of a difference was found.

3.4.3 The Null Hypothesis Is False Before Data Collection

Whether a defendant is the real perpetrator or not is a binary, discrete state. In real-world trials, there are cases where someone is actually the perpetrator. However, the differences in population means or proportions between two groups are continuous. Mathematically, the null hypothesis is almost everywhere (a.e.) covered by the alternative hypothesis. Pointing out a specific mathematical point on the real number line is scientifically absurd and nonsensical. Whether it's a difference of one-hundredth of a second in hospitalization time, a psi power of $\pi = 0.5031$, or a weight loss of one picogram, the null hypothesis is false. Testing a null hypothesis that is known to be false even before collecting data is meaningless.

As stated in Chap. 1, rejecting the null hypothesis based on statistical significance is only a necessary condition for scientific usefulness. We likened this mechanism to sending someone to the Olympics based solely on confirming they are human, without looking at their records. There is no exaggeration in this analogy. Since the null hypothesis is false even before data collection, it's too lenient as a necessary condition. Rejecting the null hypothesis is not an appropriate condition to be cleared for peer review.

3.5 Tiger at the Front Gate, Wolf at the Back Gate

When planning an experiment, it is necessary to consider both α and β. However, by convention, instead of β, we focus on the Power of the test, which is $1-\beta$. The Power of the test, being $1-\beta$, is the probability that a significant difference is detected when the null hypothesis is false (i.e., the alternative hypothesis is true). It's the probability that the test functions as a test should.

Summarizing the discussion on varying n in Eq. (3.6), we find:

- If n is too large, the Power of the test becomes too high, leading to reports that are statistically significant but scientifically meaningless. (Wolf at the back gate)
- If n is too small, the Power of the test becomes too low, burying discoveries that are scientifically meaningful but not statistically significant. (Tiger at the front gate)

So far, we have emphasized the drawbacks of the former. However, the latter is also serious. Experiments with low Power are like entering a battle with no chance of winning. They are a waste of research resources/activities. It is desirable to avoid missing a scientifically meaningful difference at the population level, due to a Type II error, especially when such a difference actually exists.

Since there are tigers at the front gate and wolves at the back gate, **you must not casually decide on n when conducting a significance test.** It is necessary to determine the right number of observations that is neither too large nor too small. This is the premise of the Power analysis by Cohen [10] (pre-analysis). With the Type II error sufficiently reduced (i.e., increased Power of the test), we test for a substantively scientifically meaningful difference with an appropriate n. The idea is to determine n in advance with a solid rationale.

3.6 Power Analysis

The state of a significance test is determined by four indicators: the probability of a Type I error α, the number of observations n, the degree of deviation from the null hypothesis, and the test power $1-\beta$. When three of these conditions are fixed, the fourth one is determined.

Assuming the other two conditions are the same, the relationship between the test power and the other three factors is as follows:

(1) Increasing α increases the test power. Decreasing α decreases the test power.
(2) Increasing n increases the test power. Decreasing n decreases the test power.
(3) Greater deviation from the null hypothesis results in higher test power. Lesser deviation from the null hypothesis results in lower test power.
Regarding (1), since α is mostly kept at 0.05, there is little room for adjustment. For (2), though there is no mathematical basis, Cohen recommends $1-\beta=0.8$. This suggests that if the alternative hypothesis is true, on average, a significant

difference will be correctly detected in 8 out of 10 trials. (3) requires careful understanding. The true state of deviation from the null hypothesis is the research result itself and is essentially unknown. Research is conducted because it is unknown. Here, it is used as the smallest deviation worth detecting for substantively scientifically significant differences. Combining (2) and (3), we get:

(4) The greater the deviation from the null hypothesis, the smaller the required n.

This leads to an important insight.

Fixing α at 0.05 and summarizing the above discussion, we obtain

$$n = f(\text{test power, degree of deviation from the null hypothesis}) \tag{3.7}$$

This shows that the number of observations can be determined as a function of two independent variables. The left side is obtained as a continuous quantity, so it is rounded up to the first decimal place to determine the minimum n required to ensure the specified test power.

3.6.1 *Determining n for Tests of Proportions*

In the "Psychic Ability Experiment," a situation where a binary event occurs "entirely by chance" was represented by the hypothesis "the population proportion is 0.5." Let's determine the n for the test of the proportion under the null hypothesis H_0 : $\pi_0 = 0.5$. What would be the appropriate value for n? In this case, the degree of deviation from the null hypothesis can be represented by the absolute difference $|\pi_0 - \pi_1|$ between π_0 and the single-point alternative hypothesis $H_1 : \pi_1$.

Table 3.2 shows the n for tests of proportions determined by Eq. (3.7). The first and second rows are the single-point alternative hypothesis $H_1 : \pi_1$. The third row is the absolute difference between the null and alternative hypotheses.

Rows 4, 5, and 6 show the number of observations for test powers of 0.7, 0.8, and 0.9, respectively. For example, with $\alpha = 0.05$, a test power of 0.8, and $\pi_1 = 0.7$, n is read as 47.

In rows 4, 5, and 6, n decreases as you move to the right, reflecting property (4) mentioned earlier. Additionally, in each column, n increases as you move down, reflecting property (2) mentioned earlier.

Table 3.2 Determining n for tests of proportions ($\alpha = 0.05$)

π_1	0.55	0.6	0.65	0.7	0.75	0.8	0.85	0.9	0.95
$\|\pi_0 - \pi_1\|$	0.45	0.4	0.35	0.3	0.25	0.2	0.15	0.1	0.05
	0.05	0.10	0.15	0.20	0.25	0.30	0.35	0.40	0.45
$1-\beta=0.7$	616	153	67	37	23	15	11	8	5
$1-\beta=0.8$	783	194	85	47	29	19	14	10	7
$1-\beta=0.9$	1048	260	114	63	39	26	18	13	9

3.6.2 *Determining the Sample Size n for the Test of Mean Differences Between Two Independent Groups*

Next, let's determine the appropriate sample size n for the test of mean differences between two independent groups, in the context of a control experiment with antibiotics. The null hypothesis $H_0 : \mu_{\text{experiment}} = \mu_{\text{control}}$ assumes no difference in the average duration of hospital stay between the two groups. The extent of deviation from this null hypothesis can be represented by a function of parameters similar to the first part of Eq. (3.2):

$$\delta_{\text{internal}} = \frac{\mu_{\text{control}} - \mu_{\text{experiment}}}{\sigma_{\text{internal}}} \tag{3.8}$$

Here, σ_{internal} is the population parameter for s_{internal}. In the case of antibiotics, this represents the difference in the average duration of hospital stays caused by the two drugs, divided by the within-group standard deviation (the average variability of hospital stay within conditions). Therefore, δ_{internal} is referred to as the **mean difference standardized by within-group standard deviation**. δ_{internal} indicates how many times the mean difference is of the within-group standard deviation. Note that Eq. (3.8) is not calculated from the data as $(\bar{x}_{\text{control}} - \bar{x}_{\text{experiment}})/s_{\text{internal}} = (9.6 - 8.1)/2.3$; please be careful about this (this is also true for $|\pi_0 - \pi_1|$ in Table 3.2).

Using the standardized mean difference, we can express a certain type of divergence between two groups that is independent of measurement units. This is because the mean difference is recalibrated in terms of the average variability within the groups. The concept of δ_{internal} when the standard deviations of the two groups are essentially the same is illustrated in Fig. 3.3. The left diagram shows $\delta_{\text{internal}} = 0.3$, the middle diagram $\delta_{\text{internal}} = 0.5$, and the right diagram $\delta_{\text{internal}} = 0.9$. Multiplying δ_{internal} by 10 allows us to interpret the mean difference in terms of the standard deviation score. A δ_{internal} of 0.3 corresponds to a difference of 3.0 in standard deviation score.

Table 3.3 shows the sample size n for the test of mean differences between two independent groups determined by Eqs. (3.7) and (3.8). The first and second rows

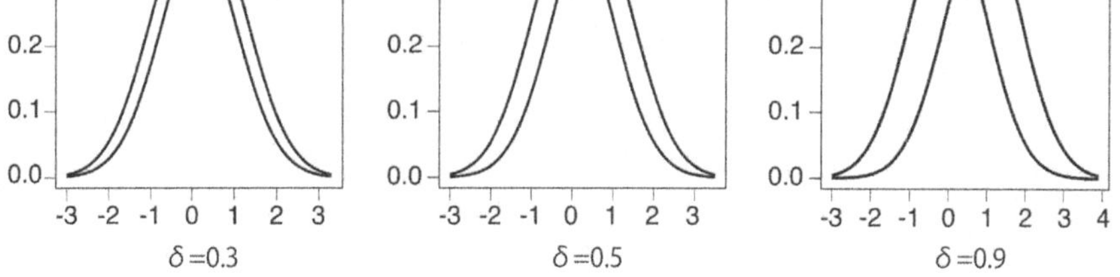

Fig. 3.3 Standardized mean difference

Table 3.3 Number of observations required for the test of mean differences between two independent groups ($\alpha = 0.05$)

δ_{within}	0.1	0.2	0.3	0.4	0.5	0.6	0.7	0.8	0.9	1.0
	– 0.1	– 0.2	– 0.3	– 0.4	– 0.5	– 0.6	– 0.7	– 0.8	– 0.9	– 1.0
$1 - \beta = 0.7$	1236	310	139	79	51	36	27	21	17	14
$1 - \beta = 0.8$	1571	394	176	100	64	45	34	26	21	17
$1 - \beta = 0.9$	2103	527	235	133	86	60	44	34	27	23

represent the mean difference standardized by within-group standard deviation. The third, fourth, and fifth rows show the number of subjects observed for test powers of 0.7, 0.8, and 0.9, respectively.

For example, with $\alpha = 0.05$, a test power of 0.8, and $\delta_{\text{within}} = 0.5$, n is 64. In rows 3, 4, and 5, n decreases as we move to the right, which is due to the property (4) mentioned earlier. In each column, n increases as we move down, reflecting property (2) mentioned earlier.

3.6.3 Determining n for the Test of Mean Differences Between Two Related Groups

Finally, let's determine the appropriate n for the test of mean differences between two related groups under the null hypothesis $H_0 : \mu_{\text{before}} = \mu_{\text{after}}$ (no difference in average weight before and after dieting). In this case, we can use the following parameter function as a measure of deviation from the null hypothesis:

$$\delta_{\text{difference}} = \frac{\mu_{\text{before}} - \mu_{\text{after}}}{\sigma_{\text{difference}}} = \frac{\mu_{\text{difference}}}{\sigma_{\text{difference}}} \tag{3.9}$$

Here, $\sigma_{\text{difference}}$ is the population parameter of $s_{\text{difference}}$. In the case of the diet problem, this formula represents the average difference in weight before and after, divided by the standard deviation of the difference scores (the average variability of weight change). Therefore, $\delta_{\text{difference}}$ is referred to as the **standard deviation-normalized mean difference of difference scores**. It's an indicator of how many times the mean weight change is of the standard deviation of difference scores. Note that Eq. (3.9) is not calculated from the data as $(\bar{x}_{\text{before}} - \bar{x}_{\text{after}})/s_{\text{difference}} = (50.57 - 47.83)/3.81$.

The concept of $\delta_{\text{difference}}$ can be understood similarly to Fig. 3.3. The left image shows $\delta_{\text{difference}} = 0.3$, the middle image shows $\delta_{\text{difference}} = 0.5$, and the right image shows $\delta_{\text{difference}} = 0.9$. You can interpret δ in these images as $\delta_{\text{difference}}$.

Table 3.4 shows the n for the test of mean differences between two related groups determined by Eqs. (3.7) and (3.9). The first and second rows are the standardized

Table 3.4 Number of observations required for the test of mean differences between two related groups ($\alpha = 0.05$)

$\delta_{\text{difference}}$	0.1	0.2	0.3	0.4	0.5	0.6	0.7	0.8	0.9	1.0
	− 0.1	− 0.2	− 0.3	− 0.4	− 0.5	− 0.6	− 0.7	− 0.8	− 0.9	− 1.0
$1 - \beta = 0.7$	620	157	71	41	27	20	15	12	10	9
$1 - \beta = 0.8$	787	199	90	52	34	24	19	15	12	10
$1 - \beta = 0.9$	1053	265	119	68	44	32	24	19	16	13

mean differences of difference scores. Rows 3, 4, and 5 show the number of observations for test powers of 0.7, 0.8, and 0.9, respectively. For example, with $\alpha = 0.05$, a test power of 0.8, and $\delta = 0.5$, n is 34.

3.7 Limitations of Power Analysis

Criticism of significance testing has paralleled its development since the 1960s. In 1988, Cohen published a textbook on **Power Analysis** [10], advocating the importance of determining a sample size, n, that is neither too large nor too small, based on sound reasoning, for significance testing. However, power analysis, with its long history, has not been universally adopted. Why is this? Are researchers incompetent? No, there are reasons for its lack of adoption. In this section, we will learn about the drawbacks and limitations of power analysis.

Firstly, many datasets in the world are not collected with power analysis in mind. Data that can be prospectively collected with a fixed n as per power analysis are rather special in the context of analysis-worthy data.

For example, cases where data are automatically collected from the web or dieting effects are studied through monitor surveys, it's often impossible to predetermine n. Most real-world data does not fit well with power analysis.

Secondly, the nature of significance testing, where "larger n leads to significance even if scientifically meaningless," is not a valid methodology for data analysis. Since data analysis is about extracting information from data, larger n should ideally be better. The occurrence of phenomena like "too much data" itself indicates a flaw in the methodology.

Bem's "Feeling the Future" study did not conduct power analysis. What if Bem had performed power analysis and predetermined an n? Given that the study aimed to demonstrate the existence of psychic abilities, any sensible peer reviewer would require a π of 0.8. Refer to Table 3.2. If we set the test power as recommended by Cohen to 0.8, Bem could have conducted only up to $n = 19$ trials, far from the actual $n = 1560$.

Power analysis has a mathematical property: "Calculating n targeting a large deviation from the null hypothesis results in a smaller appropriate n than targeting a small deviation." In everyday terms, it means, "To conduct meaningful and ambitious research, it's necessary to register a smaller n in advance." This property is correct from the perspective of significance testing. However, smaller n leads to instability in sample proportions and even the p-value itself. Essentially, by adopting power analysis, significance testing is, in effect, undermining itself. Research that is truly valuable and ambitious should involve investing resources to collect ample data and stabilize the results. Power analysis doesn't allow researchers to follow this natural inclination.

Thirdly, the deviation indicator used in power analysis may not always be a valid measure of the effect size in research. Let's explain this using the development of a one-month diet method as an example. For studies using the mean difference between two related groups and conducting power analysis, Eq. (3.9) must be used as an indicator of deviation from the null hypothesis. This indicator measures how many times the average weight loss is of the standard deviation of weight changes. Before starting the research, what $\delta_{\text{difference}}$ can be considered effective for this diet method? How should an analyst define and predetermine $\delta_{\text{difference}}$ and n?

The effective $\delta_{\text{difference}}$ is a figure based on domain-specific knowledge and is not determined statistically. The value of $\delta_{\text{difference}}$ considered valid varies among peer reviewers. However, this variability makes predetermination impossible and leaves researchers uneasy. Consequently, suppose a consensus is reached in a relevant academic society to determine n based on "$\delta_{\text{difference}} = 0.3$." This agreement would be necessary for progress in power analysis.

After data collection, let's consider a situation where the standard deviation of the difference scores among participants happened to be small, at 2 kg. This would be information revealed only after post hoc data collection in a power analysis. Consequently, the target weight loss would be just 0.6 kg ($= 2 \times 0.3$). Is such a low target truly adequate?

Conversely, if the participants are diverse and the standard deviation of the difference scores is large, say 12 kg, then the target weight loss becomes 3.6 kg. Despite both cases having the same $\delta_{\text{difference}} = 0.3$, the latter scenario poses a target six times more stringent than the former. Losing 0.6 kg in a month compared to 3.6 kg carries entirely different implications. It is strange that the true research target only becomes clear after the fact, having fixed $\delta_{\text{difference}}$ beforehand. Choosing an indicator of deviation from the null hypothesis[5] for enabling power analysis is like confusing the means with the end.

The fourth problem is an extension of the third. The indicator for expressing research effects (effect size) should prioritize demands from reality and allow for more freedom in selection. For example, from the perspective of preventing lifestyle diseases, the development goal for a diet method might need to be "a 95% probability of losing more than 2 kg in a month." The relative indicator of effect within a group, $\delta_{\text{difference}}$, does not directly connect to the goal of preventing lifestyle diseases. However, for conducting power analysis, one must select $\delta_{\text{difference}}$, making the method inconveniently restrictive.

[5] In other words, the "indicator of deviation from the null hypothesis" required by power analysis may not always be an appropriate measure of effect size for the research.

Moreover, in situations with participants of various body types, even the absolute amount of weight loss may not be a meaningful research goal. Losing 3 kg for a 100 kg person versus a 40 kg person has different meanings. In such cases, the development goal might need to be "a 95% probability of reducing obesity level BMI (Body Mass Index) by 1.0 or more in a month," considering both height and weight. Power analysis, unable to adequately reflect freely set research objectives based on real-world demands, is inconvenient in this sense. This issue will be resolved in Chap. 7.

3.8 Summary

In significance testing, the null hypothesis cannot be actively supported. For example, even if the test result is not significant, it does not justify the claim that the population means of two groups are equal. However, even professional researchers often make incorrect interpretations. This is due to a pressing scientific need to demonstrate that "there is no difference" or "there is no effect." For instance, showing that there is no difference in the condition of experimental and control groups before administering Antibiotics A and B would actively demonstrate that the controlled experiment has been conducted appropriately. The desire to show this can lead even professional researchers to misuse the tests. This is problematic. A methodology that can **actively demonstrate that two groups are substantially equal in terms of the population mean** is needed. This will be introduced in the latter part of this book.

The number of observations in significance testing should neither be too many nor too few. If n is too large, the test power becomes too high, leading to statistically significant but scientifically meaningless reports. Conversely, if n is too small, the test power becomes too low, causing scientifically meaningful findings to be statistically insignificant and overlooked. Determining n without justification is unacceptable. Justifying n for a significance test because, for instance, a surveyed class happened to have 40 students, all 100 new students in a basic psychology experiment needed to participate, or 53 participants were gathered in a lab reserved for three days, does not suffice.

When writing psychology papers, one of the most utilized references is the APA Publication Manual [11]. Edited by the American Psychological Association, the APA Manual is referenced not only in psychology but in many scientific fields as a guidebook for writing papers. This manual specifies that authors should include in the 'Methods' section of their papers the method used for determining the intended sample size.[6,7] Despite the clear directive to conduct power analysis, many researchers,

---

[6] Sample size refers to the number of observations, n. It is a term unique to significance testing for deriving a sample distribution.

[7] The original text states: "State how this intended sample size was determined (e.g., analysis of power or precision)."

including psychologists, often fail to comply with this when writing papers and reports. They leave it to chance or arbitrarily decide on n, which is disingenuous. Such practices are far from conducive to research reproducibility.

If one is to perform significance testing, it is necessary to determine the appropriate number of observations using power analysis, ensuring it is not too many or too few (neither prey to the tiger nor the wolf). Cohen emphasizes in his textbook that n should not be too large or too small. When conducting significance testing, researchers must provide a rationale for their chosen number of observations based on power analysis in their papers or reports. However, power analysis comes with various shortcomings.

If power analysis were truly a useful method, one might expect standard statistical textbooks over its long history to have evolved to consistently pair each testing method with a method for determining n. Indeed, the policy of predetermining n in itself is inconvenient and restrictive. Specific solutions that do not require predetermined n will be introduced in Part II.

Column: The Multiple Comparisons Problem

Here, we address the **multiple comparisons problem**, a serious flaw in significance testing, and introduce the concept of **intention**, which will be discussed more thoroughly in the next chapter.

The Multiple Comparisons Problem by Levels

Until now, we have discussed cases in the context of dieting or antibiotics where there was one treatment and one control. Here, we consider cases where there are multiple states corresponding to the treatment. For instance, let's consider whether the average scores on a survey question differ based on "blood type." Here, a collection of subgroups like "blood type" is referred to as **attributes**. Individual states of attributes, such as A, B, O, and AB blood types, are called **levels**.

Imagine a survey question stating, "I think I am the organizer among my friends." Respondents answer using a **multiple-choice method** (in this case, a seven-choice method) ranging from "very applicable" to "not applicable at all." Suppose the average score is higher for type O than for type AB, and an independent two-group mean difference test results in $p = 0.04$. Can we say it is significant at the 5% level?

If the analyst had a prior academic conviction that "leadership qualities differ specifically between type O and AB," and conducted the test with the intention of comparing only these two blood types without considering others, then we can conclude it is significant at the 5% level. However, if the test was conducted with the academic conviction that "leadership qualities differ by blood type," and all possible pairings were compared six times,[8] the scenario changes. If even one of these six tests results in a Type I error, the conclusion would be that "leadership is significantly influenced by blood type." The probability of this error would be higher than 5%. This is called α **inflation**.

The group of methods that adjust the p-value to prevent α inflation during multiple tests is collectively referred to as **multiple comparison methods**. For instance, in one such method, the **Bonferroni method**, a difference is considered significant at the α level if $p < (\alpha/\text{number of tests})$. In the aforementioned example, with four levels and six combinations of tests, the number of tests is 6, and $p = 0.04 > (0.05/6) = 0.008\dot{3}$, so it is not significant at the 5% level.

Now, a bizarre property of significance testing is revealed. Despite using the same data, the determination of significance changes based on the intention. The bizarre aspect here is that "since intention is invisible, peer reviewers and readers cannot discern which intention the analyst had." This issue is profoundly significant. However, there is a solution. Based on the principle that the burden of proving academic value rests with the analyst, it would be fair to assume the latter intention. Thus, when there are three or more levels, applying a multiple comparison method becomes the principal analytical procedure. But even with adjustments, this is as far as it goes.

The Multiple Comparisons Problem by Attributes

In a previous survey, there was a question, "I think I am a respectable person." Among men, type O had a higher sample mean than type AB, and testing yielded $p = 0.005$. Since $p < 0.008\dot{3}$, can we say it's significant at the 5% level? If the analyst had a prior conviction that "among men, self-esteem differs by blood type" and tested only men with that intention, then we can conclude it's significant at the 5% level. However, if the test was conducted with the intention that "self-esteem differs by blood type in either men or women," the story changes. With 12 tests in total for men and women, making at least one Type I error would have led to the conclusion that "self-esteem is significantly influenced by blood type." Therefore, it must meet $p < (0.05/12) = 0.0041\dot{6}$ to be significant at the 5% level, which it does not.

Among women, type O also had a higher sample mean than type AB, and testing yielded $p = 0.001$. Since $p < 0.0041\dot{6}$, can we say it's significant at the 5% level? If the analyst had a conviction that "self-esteem differs by blood type in terms of gender" and tested only within genders, then we can conclude it's significant at the 5% level. However, if the test was conducted with the intention to find differences in self-esteem by blood type in any attribute and "gender" was chosen by chance, it's a different matter. The survey included attributes like "gender (2)," "age group (6)," "income group (6)," "prefecture (47)," and "occupation group (39)." There

[8] The number of combinations for comparing two out of m items is $m \times (m-1)/2$. For $m = 4$, there are 6 combinations.

were 100 comparison subjects, resulting in 600 tests. This hyperinflation means that $p < (0.05/600) = 0.00008\dot{3}$ is needed to be significant at the 5% level, which it does not meet.

Testing with attributes and levels, as described, assumes various testing intentions, each with a different degree of α inflation. Since intention is invisible and the burden of proof of academic value lies with the analyst, always using $p < 0.00008\dot{3}$ as the criterion for a 5% level test is a fair approach. However, with an increased number of tests, it becomes difficult to find significant differences. Fair and correct application of multiple comparison methods often results in no significant differences anywhere, negating the purpose of conducting the survey. Therefore, it's common practice to compromise and apply multiple comparisons only within levels of each attribute, as a sort of excuse that "correction has been made." But this method does not prevent α inflation between attributes, contributing to the lack of research reproducibility.

The same applies to tests of proportions. Let's consider a combination of two attributes. In the survey, there was a question, "Do you agree with Bill A currently under discussion in the parliament?" answered with "Yes/No." The sample approval rate was higher for men in Prefecture A than in Prefecture B, resulting in $p = 0.04$. Naturally, this is not significant at the 5% level. Unlike the previous examples, here comparisons are made by combining two types of attributes. The number of tests[9] is 3097, requiring $p < (0.05/3097) = 0.00001614$ to be considered significant at the 5% level. Though considering combinations of two attributes is often done, fair and correct p-value correction is rarely performed. However, it's harsh to blame the analysts. The theoretical structure of significance testing should be considered as inherently flawed due to the multiple comparisons problem.

The Multiple Comparisons Problem by Criterion Variables

The variable that forms the objective of the research is referred to as the **criterion variable.**[10] Until now, we have discussed situations involving a single criterion variable, such as "weight" or "hospital stay duration," in problems related to dieting and antibiotics. If a survey contains 60 questions, each question is a criterion variable. How should we adjust the p-value in this case? If we consider two attributes with the intention of finding differences by "blood type," the criterion becomes $p < (0.05/(60 \times 3097 \times 6)) = 0.000000045$ for declaring a significant difference at the 5% level. In the modern era of advanced PCs, performing this number of tests is easily achievable with just a few lines of code.

---

[9] $3097 = 2 \times 6 + 2 \times 6 + 2 \times 47 + 2 \times 39 + 6 \times 6 + 6 \times 47 + 6 \times 39 + 6 \times 47 + 6 \times 39 + 47 \times 39$. There are 10 combinations for selecting two levels from five levels. The number of comparisons for each pair of attributes is the product of their levels. Sum the number of comparisons for all 10 pairs. "Blood type" is not considered.

[10] The criterion variable is also known as the **dependent variable**, **explanatory variable**, **outcome**, or **endpoint**.

In cancer treatment, for example, there are eight criterion variables like "overall survival," "progression-free survival," "disease-free survival," "tumor shrinkage rate," "response rate," "5-year recurrence," "quality of life," and "treatment cost." How should we adjust the p-value when testing for differences between four treatment methods (A, B, C, D)? If there was an intention to write a paper if a significant difference is found in any criterion variable between any treatment methods, the number of tests would be 48 (8×6). Therefore, a difference is considered significant at the 5% level if $p < (0.05/48) = 0.00104$. If attributes like "presence of close relatives with the same disease (2)" or "gender (2)" ... are also considered, the number of tests increases exponentially, as explained in the previous section.

Snooping in Multiple Comparison Methods

There are multiple comparison methods besides Bonferroni's method. Bonferroni's method, known for its wide applicability and simplicity, was used in the earlier explanation as a representative example of multiple comparison methods. All multiple comparison methods become more stringent in evaluating p-values as the number of tests increases, and they all share the fundamental flaw of depending on the analyst's intention. Bonferroni's method is known for its low power. Given the same degree of deviation from the null hypothesis, n, and α, there are methods with higher testing power. Methods such as **Scheffé's method**, **Dunnett's method**, **Tukey's method**, **Williams' method**, **Holm's method**, **Steel's method**, and others offer analysts a more advantageous approach due to their higher power.

However, each multiple comparison method is tailored to specific intentions and naturally yields different results. Thus, analysts might apply methods arbitrarily and then write papers as if they had originally intended the method that provided favorable results. This practice is known as **snooping** in multiple comparison methods. It contributes to the inflation of α and leads to the lack of reproducibility in research. The proliferation of testing methods is also one of the weaknesses in the theoretical framework of significance testing.

Levels, attributes, and criterion variables are explicitly stated in research papers, leaving evidence of any rule-breaking in multiple comparisons. However, the dependency of results on "invisible intentions" is in itself a significant flaw of the p-value. Yet, the issues discussed here can be considered relatively minor in the sense that they leave evidence of misconduct. The **multiple comparisons problem due to timing**, discussed in the next chapter, possesses a more despairing nature as it leaves no trace of rule violation, creating more severe problems for reproducibility in research. As we delve into this perfect crime without evidence, the mystery intensifies and reaches its climax.

Chapter 4
The Zombie Problem

Does the Fact of "Deciding in Advance" Guarantee Permanence Post-Hoc?

4.1 Testing Every Time

There are people who, for health management, measure their weight daily and calculate their obesity level. Meteorologists, likely, check the temperature every day. Observing, measuring, and analyzing in response to changes in the subject of interest and continuously monitoring it is a natural part of scientific activity. It represents a sincere scientific attitude of not wanting to miss any changes in the subject. Scientists who do not monitor their research subjects are, in fact, lazy. Of course, there is no malice in this. But what about the following behavior?

> Compare the non-cure rate (unknown) of a new treatment method with the non-cure rate (known to be 0.5) of an old treatment method. At milestones of increasing case numbers, results of significance tests were presented several times at interim announcement meetings. Each time, the supervising professor praised it as "promising," so the treatment was continued with confidence. Cases were observed and measured in detail, and a preliminary binomial test of the ratio was conducted with each additional case, continuing the treatment while watching the situation.
>
> One day, the long-awaited significant difference was observed. The researcher was overjoyed, thinking that years of hard work had paid off. "After

H. Toyoda, *Statistical Significance and the PHC Curve*,
https://doi.org/10.1007/978-981-97-7748-8_4

treating n people, unfortunately, x did not recover. However, the sample non-cure rate x/n is statistically significantly lower than the non-cure rate of the old treatment method, as shown by the binomial test," was recorded in the doctoral thesis as fact and final test result.

While increasing the number of observation subjects and conducting multiple tests, you must not report only the final test result in papers or reports. If you want to do that, you need to describe in your paper "how many times and at what points you conducted preliminary tests before the final test" and report the **adjusted *p*-value** as per the **sequential test** (sequential test) [12] or **adaptive design** (adaptive design) [13] methods. Reporting as if the final result was the first binomial test without adjustment increases the probability of committing a Type I error (judging a treatment with no effect as effective).

In situations where performing numerous tests can lead to making some claims if any of them is significant, an inflation of α occurs. This is commonly referred to as the **multiplicity problem**. The disadvantages caused by multiplicity problems arising from levels, attributes, and criterion variables have been discussed in a column. Here, we address the multiplicity problem that arises over time. The research process described above violates the rules of significance testing concerning **temporal multiplicity**.

The APA Publication Manual [11] clearly states that the 'Methods' section of a paper should include "The original text is as follows: If interim analysis and stopping rules were used to modify the desired sample size, describe the methodology and results of applying that methodology." It means you must report the strictly adjusted p-value using methods like sequential testing or adaptive design. However, even psychologists with their own writing manuals do not adhere to this. Could there be such deceit?

S Teacher! I had to speak out because what you are saying is just too terrible.
T What about it is terrible?
S It is an undeniable fact that $n - x$ people were cured after treating n people. And it is also a fact that a significant difference was obtained. Then, what analyses were done previously have nothing to do with it! The fact that today's weight is 49 kg and the temperature is 23° is not negated by the weight and temperature measurements up to yesterday. Similarly, the fact that it became "significantly different" today is not negated.
T But it's still not acceptable.
S First of all, "If you conduct multiple significance tests midway, you should not only describe the final test result," is not written in the psychological statistics textbooks I have. It's not in introductory testing textbooks either.
T It's true that it is not explained in introductory statistical textbooks currently published, not just in psychology. But the APA manual does state to "write the name of the method and its application results." Monitoring p-values while increasing data and writing a paper when a significant difference is obtained is out of the question.

If you are going to teach significance testing, you should explain the principle of α inflation in sequential testing and methods for adjusting p-values. Serious harm occurs, so you absolutely cannot omit it, even in elementary textbooks, if you recommend the use of significance testing.[1]

S Having a great supervising professor who plans interim announcement meetings so many times is wonderful. You can't have an interim announcement meeting without some analysis midway. Are you saying not to have interim announcement meetings? Without some analysis midway to get a sense of direction, I would be too scared and anxious to complete the research.

T I'm not saying don't do it. Rather, you should. The supervising professor should record how many times tests were conducted and adjust the p-values accordingly. Having interim announcement meetings is more careful guidance. If they don't make adjustments, that itself is negligence on the part of the supervising professor. They may be accused of stealing their salary.

S But the more analyses you do at interim announcements, the more strictly the final test's p-value will be evaluated, right? Since the researcher mentioned earlier finally obtained a significant difference one day, if corrected, they couldn't write a paper, right? Then I don't want to do interim announcements ... or rather, the act of testing every case at home in the first place leaves no evidence at all!

T That's right. Unlike the case with level, attribute, and criterion variables, the multiplicity problem of timing is dark because there is no evidence of cheating left.

S In the case of joint research, you don't know what other people are doing. Are you going to monitor each other's lives with security cameras to not perform sequential significance testing? Even if you monitor, if you test with paper and pencil in a camera-less bathroom and then throw away the paper, it's as if you didn't do it.

T What you say makes intuitive sense. That's why the rules of significance testing regarding temporal multiplicity are broken without malice. Reviewers only demand the final result of $p < 0.05$, so many researchers don't adjust p-values. But if you break this rule, the probability of Type I errors will definitely inflate. The true risk rate of a test with an apparent risk rate of α 5.

S And all without any malice.

T Actually ... it's not completely without malice. It's quite widely known as a secret to "succeed" in research that if you test occasionally and continue to collect data, it becomes easier to obtain significance. This isn't written in textbooks, though.

S Really?

T By "using" this principle, researchers may discover significant differences in places they don't academically believe have any. Due to the harsh competition between researchers, many of them may be published. It's the result of α inflation, so many of them will not be reproducible.

---

[1] This book does not recommend the use of significance testing, so we do not explain sequential testing or adaptive design. Those who continue to use significance testing after reading this book should study the specialized textbooks listed at the end. Those who are about to write textbooks using significance testing must definitely explain sequential testing and adaptive design.

Researchers who use significance testing as their own tool have the nature of the tool ingrained in them. They may vaguely realize this taboo, even if not theoretically. But it's not written in textbooks that "you shouldn't do it," and it's a "skill" that unfairly benefits them ... Maybe some researchers are pretending not to notice and wiping their mouths.

S I understand what you're saying, Teacher. But I have zero sense of satisfaction.

T Then let's carefully explain the nature of significance testing concerning multiplicity to make you understand.

4.2 The Sample Distribution Relies on Unrecorded Intent

Why does the problem of multiplicity arise in significance testing? It's because the sample distribution of the test statistic actually depends on the **intent** of the analyst [14]. The sample distribution represents the distribution of the statistic calculated from data that was not actually observed, imagined as if countless repeated experiments were conducted with a specific intent.

The intent of the analyst is not visible. The analyst might not even be fully aware of it. Moreover, clarifying the intent held during the analysis is a challenging task, as it's often not evident retrospectively. Let's consider the following small numerical example.

Investigate whether the non-cure rate (unknown) of a new treatment method is less than the non-cure rate (0.5) of an old treatment method. Out of 24 patients, 7 did not recover. The non-cure rate is 0.29 (= 7/24). Is this sample ratio of 0.29 significantly smaller than 0.5?

This is all the data. However, significance testing cannot be specified with data alone. We need to clearly understand the intent of the researcher regarding how this data was collected. Let's examine four typical intents.[2]

4.2.1 Intent 1: Fixing n and Observing x

The researcher began the clinical experiment with the intention of "treating 24 patients." As a result, 7 patients did not recover.

In this case, for significance testing, we assume the null hypothesis $H_0 : \pi = 0.5$ to be true, and imagine infinitely repeating trials (patient numbers) of 24. Furthermore,

[2] A fifth intent is also possible: "I was just testing according to what's written in the textbook, and I never thought that intent was necessary for testing." In fact, this fifth intent might be the most common. (Tears).

for each of these repeated trials, imagine checking and recording the number of non-recoveries x. For example, in one trial, 4 patients didn't recover, in another, 8 didn't recover ... and so on. The imagined distribution of the number of non-recoveries is called the sample distribution of x, which becomes a binomial distribution with $n = 24$. This is the **binomial test** we studied in Chap. 1. Intent 1 is the initial setting learned in elementary statistical textbooks. Hereafter, Intent 1 may sometimes be referred to as textbook Intent 1.

Under this intent, if we perform significance testing, the one-sided probability of having 7 or fewer non-recoveries is 0.032. This probability is greater than 0.025, so considering the upper side for x, the test result is determined to be "no significant difference" at the 5% level.

The concern here is whether the clinical experiment really started with the intention of "treating 24 patients." This less visible intent can fundamentally influence the results. We will demonstrate this with the next Intent 2.

4.2.2 *Intent 2: Fixing x and Observing n*

The researcher began treatment with the intention of "testing the treatment effect after failing to cure 7 patients." In the clinical experiment, the 24th patient became the 7th non-recovered patient. The treatment results are exactly the same as in textbook Intent 1.

In significance testing, we assume the null hypothesis $H_0 : \pi = 0.5$ to be true, and imagine infinitely repeating the exact same experiment. But unlike Intent 1, we imagine repeating trials where treatment continues until there are 7 non-recoveries. Furthermore, for each of these repeated trials, imagine checking and recording the number of patients n. For example, in one trial, treatment was given to 23 patients, in another, to 27 patients ... and so on. While x was distributed in Intent 1, in Intent 2, n is distributed.

The imagined distribution of the number of recoveries $n - x$ at this point is known to follow a **negative binomial distribution**

$$\begin{aligned}&\text{Negative Binomial Distribution}_x(n - x|\pi)\\&\quad = \text{Negative Binomial Distribution}_{\text{Non-Recoveries}}(\text{Recoveries}|\pi)\end{aligned} \tag{4.1}$$

Figure 4.1 shows

$$\text{Negative Binomial Distribution}_{x=7}(n - 7|\pi = 0.5) \tag{4.2}$$

with n on the horizontal axis. The lower limit of the number of patients n is, of course, 7. Since the negative binomial distribution has no upper limit, we show it up to 30 patients. This is the sample distribution of the number of patients n (the test statistic).

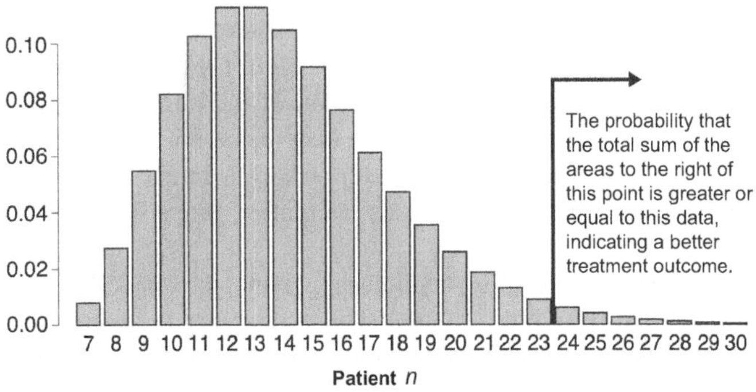

Fig. 4.1 Sample distribution of patient numbers by negative binomial distribution

Under the true null hypothesis, the probability of having better treatment results than the given data is the probability that the number of patients n is 24 or more (non-recovery rate is 7/24 or less). It is the sum of the areas of all bars to the right of 24 in Fig. 4.1. However, since the negative binomial distribution has no upper limit, it cannot be summed up. We use the fact that the sum of a probability distribution is 1. What we do is sum the areas of the bars from $n = 7$ to 23 and subtract that from 1.

The calculated one-sided probability on the upper side turned out to be 0.017. This probability is smaller than 0.025, so considering the lower side for n, the test result is determined to be "significant difference" at the 5% level.

Now, something unbelievable has happened. Analyzing the same treatment data ($n = 24, x = 7$), the significance test results vary depending on the intent—sometimes significant, sometimes not. The unbelievable part is that **a third party cannot know the true intent of the analyst**.

Even more problematic is that both Intent 1 and Intent 2 are, after all, unnatural intents set to make the mathematical derivation of the sample distribution easier. In experiments/surveys conducted with a sincere motivation, it's more natural to fall under the following Intent 3, rather than Intent 1 or 2.

4.2.3 Intent 3: Fixing Research Resources, Observing n, and Then x

The researcher began the experiment intending to "test using the treatment results of patients who visit within a year." In the clinical experiment, 24 patients visited and received treatment within a year, and 7 of them did not recover. The treatment results are the same as in Intent 1 and Intent 2.

Intent 3-based testing is most commonly seen in real-world scenarios.

- After managing to gather as many subjects as possible during the 10 days the laboratory was reserved, n subjects were secured, and a positive response was obtained from x subjects.
- Intending to let all new students experience a psychological experiment educationally every year, this year n subjects were targeted, and a positive response was obtained from x subjects.
- In an effort to meet the deadline for a thesis, as many survey forms as possible were distributed, resulting in n respondents, of whom x supported the policy in question.
- During the survey period for traffic safety, n traffic accidents occurred, and in x of those, seat belts were not worn.
- After recruiting experience monitors for a diet method for a week, n applicants participated, and x of them lost weight.

These are also examples of Intent 3. Hereafter, Intent 3 may sometimes be referred to as (research) resource-fixed Intent 3. As we can see, most natural research falls under Intent 3. As shown earlier, intent profoundly affects the results of the test. Therefore, it is not feasible to forcibly argue that the situation was textbook Intent 1. Textbook Intent 1 is not an approximation of resource-fixed Intent 3. The binomial test found in elementary textbooks is, in reality, hardly applicable.

In significance testing, we assume the null hypothesis $H_0 : \pi = 0.5$ to be true, and imagine infinitely repeating the exact same experiment. So, under resource-fixed Intent 3, we imagine repeating trials with a research period of one year. In some trials, only 23 patients might visit in a year. In other trials, as many as 26 patients might visit, and so on. Furthermore, for each of these imagined trials, imagine checking and recording the number of non-recoveries. The distribution imagined from these nonexistent realities is the sample distribution of non-recovery numbers under resource-fixed Intent 3.

Under resource-fixed Intent 3, not only x but also n becomes a random variable, and n is known to follow a **categorical distribution**. The categorical distribution is, for example,

$$p(n = 23) = 0.2,\ p(n = 24) = 0.3,\ p(n = 25) = 0.3,\ p(n = 26) = 0.2$$

is expressed as a list of pairs of realized numbers and their associated probabilities (summing to 1). It's a type of probability distribution. In this case, the sample distribution of x becomes a **mixed binomial distribution**, which is synthesized by weighting with the categorical distribution of n and then marginalized.

Can we calculate the p-value using the mixed binomial distribution as before? No, we cannot. This is because the categorical distribution of n is unknown. The sample distribution of x in an experiment based on resource-fixed Intent 3 is at least not that of a binomial test. The true sample distribution is unknown. The tests for proportions found in elementary textbooks are, in reality, not feasible in the most commonly implemented resource-fixed Intent 3 scenarios.

4.2.4 *Intent 4: Increasing n While Monitoring x and p-Values*

Now, we are ready. Is the example at the beginning of this chapter really breaking the rules? The researcher began the experiment with the intention of "increasing the number of observation subjects, conducting binomial tests for each case, monitoring p-values, and if a significant difference is observed, writing only the result of the last binomial test in the paper." Hereafter, this Intent may be referred to as p-value monitoring Intent 4.

In significance testing, we assume the null hypothesis $H_0 : \pi = 0.5$ to be true and imagine infinitely repeating the same experiment. Of course, the intent includes the schedule for calculating the test statistic. In this case, imagine repeatedly conducting tests, and if even once the p-value falls below 0.05, the experiment ends. The risk of committing a Type I error in one test was 0.05. If we can observe a significant difference in just one of the many tests, it seems the true risk rate would be greater than 0.05.

First, we assume the null hypothesis to be true. If the p-value falls below 0.05, it means a Type I error has occurred. If one in twenty experiments allows for a paper to be written (albeit incorrectly), it would be considered a correct significance test in that sense. If more papers can be written, it implies that α has inflated unfairly. The true risk rate is confirmed to be greater than 0.05.

The experimental trials considered for the sample distribution are imaginary. However, such experiments can be realized within a computer. This is called **computer simulation**. For example, the procedure would be as follows:

1. Set an upper limit for the number of patients. For instance, suppose it's a common disease, and on average, one patient is treated per day. Let's assume that if no significant difference is found in a few years, the thesis will be abandoned.
2. Data is represented as a binary ratio of 1 or 0, with a mother ratio $\pi = 0.5$. 1 indicates recovery, 0 non-recovery. In a computer, unlike the real world, it's possible to uphold the null hypothesis.[3]
3. From the third case, perform a binomial test for each additional case. If the sample ratio exceeds 0.5 and $p < 0.05$, write the thesis. If this condition is never observed up to the patient limit, do not write the thesis.
4. Repeat this experiment 10,000 times.

[3] Since **pseudo-random numbers** are used in computers, the null hypothesis isn't precisely true, but it's close enough for practical judgment.

5. The ratio of theses written in these 10,000 experiments is the estimate of the **true risk rate** of this significance test. Compare the **apparent risk rate** of 5% with the true risk rate.

The true risk rate for a two-year effort was 0.20. For a three-year effort, the true risk rate was 0.22. Reviewers think they are testing at a risk rate of 0.05, but in reality, it is about four times that. Even a completely stagnant "new" treatment method gets a $p < 0.05$ in about one in five cases. Increasing the number of patients can further raise the true risk rate.

This simulation is for when the null hypothesis is true. However, in the world of science, the null hypothesis is false. As mentioned in the chapter on God's Invisible Hand, medically meaningless treatments that are strictly false under the null hypothesis become the subject of discussion.[4] In this case, continuing testing while increasing cases makes it surprisingly easy to observe significant differences. For example, using Bem's example with a mother ratio $\pi = 0.53$, the true risk rate for a two-year effort was 0.61. For three years, the true risk rate was 0.73. Even academically meaningless, about 60–70.

Allowing p-value monitoring Intent 4 makes statistically significant but scientifically meaningless results even easier to publish. However, in many cases, the analyst has no malice. This is because basic statistical education on temporal multiplicity is not currently provided. It is conspicuously absent. Intermediate presentations within the research lab, and even more so, preliminary tests on one's own desk leave no evidence, and whether or not a preliminary test was conducted is completely unrelated to treatment results. Therefore, it's difficult to realize the devastating impact this monitoring behavior has on the final p-value.

4.3 The Zombie Problem in Significance Testing

Recall the significance test for the difference between two independent groups introduced in Chap. 3. The sample distribution presented there was imagined based on countless repetitions of experiments with observation numbers $n_1 = 24$, $n_2 = 26$, and calculating the test statistic z according to Eq. (3.2) for each experiment. This represents a sample distribution generated under textbook Intent 1.

However, in actual research scenarios for estimating the difference between two groups, it is more natural to analyze based on resource-fixed Intent 3, where the research period (or resources) is fixed, and the observation numbers are a result of that. If similar repeated experiments are conducted over a fixed research period, the observation numbers n_1, n_2 themselves will distribute. The numbers $n_1 = 24$, $n_2 = 26$ are not the result of a deliberate collection but rather the number of patients treated during the research period, revealing the true intent. In the case of resource-fixed Intent 3, since the distribution of observation numbers is unknown, the original

[4] Just because the ratio is medically meaningless doesn't mean "the observed significance is correct because the null hypothesis is precisely false."

sample distribution of z[5] is also unknown. Testing cannot be conducted. However, it is often analyzed by masquerading as textbook Intent 1.

Intent 4 involved increasing observation subjects while repeating significance tests and monitoring p-values. The same might be done in testing the difference between two groups, reporting only the final result. The apparent final output of tests under resource-fixed Intent 3 or p-value monitoring Intent 4 is indistinguishable from that under textbook Intent 1. However, they impair research reproducibility.

In recent years, as part of **open science**, data sharing is recommended. However, even if data is published, the intents 1, 3, and 4 cannot be distinguished from each other post-publication.

4.3.1 Philosophical Zombies and Zombie-Like Tests (Definition of Terms)

In philosophy, there is a thought experiment against physicalism involving "**philosophical zombies**," which are indistinguishable from humans in appearance, function, and behavior.[6] It's a bit scary to imagine zombies, lacking consciousness but indistinguishable from us, lurking around. But "philosophical zombies" are just philosophical contemplations and won't attack us; in fact, they don't exist, so rest assured.

Tests analyzed under resource-fixed Intent 3 or p-value monitoring Intent 4, while harboring these intents but disguised as textbook Intent 1, resemble philosophical zombies in that they are indistinguishable in appearance. Therefore, henceforth, these will be referred to as **zombie-like tests** (or simply **zombies**). In contrast, tests properly conducted under textbook Intent 1 will be referred to as **Intent 1 humans** (or simply **humans**). Although philosophical zombies cause no harm, zombie-like tests devour the precious fruit of research reproducibility. This issue is referred to as the **zombie problem**. Zombies based on resource-fixed Intent 3 and p-value monitoring Intent 4 will be specifically referred to as **Intent 3 zombies** and **Intent 4 zombies**, respectively. Zombies are probably much more numerous than humans and lurk around us, hidden in papers.

4.3.2 Eradicating Zombies

Just as presenting a valid ticket at a concert entrance doesn't suffice if one's name wasn't pre-registered—making it impossible to determine post-facto whether the ticket was obtained through scalping—so too is it impossible to discern post-

[5] If the categorical distribution of observation numbers is given, z follows a **mixed *t* distribution.**

[6] Chalmers D. (1996). The Conscious Mind: In Search of a Fundamental Theory. Oxford University Press.

publication the intent behind a paper's p-value if n wasn't **pre-registered**. Even if raw data is made public later, it doesn't distinguish between humans and zombies. This is the crux of the problem. As a result, academic journals requiring pre-registration of effect sizes and n have been increasing recently.

To avoid **zombie infection**, one must perform power analysis to determine a justified number of observation subjects before collecting data, fixing it to the "Intent 1 human" as explained in textbooks. Zombies that have been infected should not be allowed to masquerade as "humans," and it's also crucial for uninfected humans to prove that they are not zombies. To this end, it is necessary to pre-register the determined number of observation subjects with a third party. This third party must guarantee that the registration was made in advance. While concert organizers only need to ensure this until the end of the concert, academic papers require perpetual assurance. Thus, **eradicating zombies** incurs a significant cost.

Moreover, this troublesome property of temporal multiplicity is not limited to binomial tests of proportions and tests of differences in means. Numerous tests not appearing in this book share the same property. If significance testing continues to be used as a standard in peer review, fundamental rewriting or modification of elementary statistical education is necessary. Furthermore, since it's an inherent flaw of the methodology of significance testing, guaranteeing it solely through peer-reviewed papers is insufficient. As long as significance testing is used, whether in classroom exercises, theses, or conference presentations, it's necessary for a third party to perpetually guarantee what was predetermined.

4.4 Confidence Intervals Do Not Save Significance Testing

So far, we have discussed the problems associated with significance testing. But what about **confidence intervals**? For example, the 95% confidence interval for the mother ratio in Bem's precognition experiment was [0.506, 0.556]. Although the probability of the mother ratio being included in the 95% confidence interval is not 95%, it is practically useful. Since the burden of proving the existence of precognition falls on the author, evaluating the lower limit yields 0.506, which means only hitting correctly 6 times out of a thousand more than chance. It's greater than 0.5 but can be deemed meaningless. Moreover, confidence intervals narrow as n increases, allowing for more certain judgments. We didn't mention this in the book, but confidence intervals for the difference in means or standardized differences in means can also be calculated. It seems that evaluating confidence intervals in a substantively scientific manner could avoid the invisible hand of God.

But does replacing significance testing with confidence intervals solve the problems we've discussed? Unfortunately, no. The problems are not solved at all. For instance, the definition of a 95% confidence interval involved continuously conducting 5% level tests while changing the null hypothesis and identifying the interval where the null hypothesis is not rejected. This means that the original confidence intervals also vary depending on the intent. In situations where the risk rate is not correctly assessed, confidence intervals are not accurate either. For example, confidence intervals derived only from the final test results in sequential testing become **inappropriately narrow confidence intervals** advantageous to the analyst, not the **original confidence intervals**. Since they are derived solely on the principle of testing, confidence intervals inherit the flaws of significance testing.

Bem's test might have been conducted with p-value monitoring Intent 4, constantly testing and adjusting n as it went. The original confidence interval might have been much wider, possibly including 0.5. The published confidence interval in the paper might have unfairly favored Bem by being too narrow. Without prior registration, they are indistinguishable. Confidence intervals derived from intents other than textbook Intent 1 are also **zombie-like confidence intervals**, indistinguishable from the original and lurking undetected in papers.

4.5 Pre-registration System

In the previous chapter, we mentioned the need to show the rationale for choosing a certain number of observation subjects in papers involving significance testing. However, this chapter's considerations show that this alone is insufficient. After obtaining a significant difference through sequential testing, it's possible to retrospectively justify the n at that point with a power analysis in the paper. Since it's required by submission guidelines, researchers naturally do this. Such invisible intent-dependent problems remain in significance testing.

An effective system against the invisible hand of God, multiplicity problems, and zombie problems in significance testing is the **pre-registration** system. For research involving significance testing, the following **registration report** must be submitted before starting the research:

(1) Explicitly state the deviation from the null hypothesis that is considered substantively scientific.
(2) Determine the number of observation subjects n based on the deviation, significance level, and test power.
(3) Declare the name of the tests and multiple comparison methods, and the method for adjusting p-values.

This system ensures proof of registration before the start of research. In my field of expertise, many journals under APA (American Psychological Association) and BPS (British Psychological Society) have implemented the pre-registration system. In the field of medical research, it's called the **clinical trial registration** system[7] and is already operational in our country. Academic journals adopting pre-registration are continuously increasing.

[7] Japan Medical Association Clinical Trials Promotion Center
https://dbcentre3.jmacct.med.or.jp/jmactr/.

4.5.1 *Benefits of Pre-registration*

For papers with pre-registered information as described in (1) and (2), the problem of God's invisible hand does not arise. For example, a diet method that only leads to an average loss of 100 g over a month (though there is a Type I error) becomes extremely difficult to publish. The test power is low for the predefined n in a good way compared to the common-sense effectiveness of a diet. Thus, papers that are technically false under the null hypothesis but scientifically meaningless can be effectively excluded.

Pre-registration also works against multiplicity problems related to levels, attributes, and criterion variables. If the method for adjusting p-values is registered in (3), you cannot just try a variety of multiple comparisons and write a paper as if those results were intended from the start.[8]

It is also effective against the zombie problem. Significance testing depends on intent and cannot be specified with data alone. However, if you pre-register, you can fix it to Intent 1 (or correctly to Intent 4 if the **test schedule** is registered in (3)). Since n is fixed by pre-registration, you cannot end the research abruptly when a significant difference is incidentally observed after testing with each added subject. Nor can you add extra data to achieve significance if it's not significant for the pre-registered n. Pre-registration, in this sense, eliminates the need to monitor researchers' lives with surveillance cameras.

A question arises here: what if the data was already collected at the point of pre-registration? If one can conduct as many preliminary tests and multiple comparisons as desired, find a significant difference, then retrospectively align the power analysis and claim "we're about to collect data" for a post-hoc "pre-registration," the system wouldn't work. No complete method has been found to prevent this issue.

To mitigate this, peer review of the content pre-registered can be effective. Peer review rarely gives a free pass and usually demands revisions. This reduces the motivation for maliciously pre-collecting data and then "pre-registering" it post-hoc. Also, if the power analysis or plan for p-value adjustments is incorrect, it defeats the purpose. Therefore, pre-review of the research plan before data collection is important for verifying its correctness.

4.5.2 *Problems with Pre-registration in Significance Testing*

Pre-registration related to significance testing comes with its own set of issues.

Firstly, securing reviewers for pre-registration is extremely challenging. Reviewing pre-registration content (1) and (2) requires deep knowledge of power analysis. It might be manageable for just the binomial tests and tests of differences between two groups introduced in this book, but what about **analysis of contingency tables**,

[8] However, if the range of p-value adjustments for levels, attributes, and criterion variables is extended to the entire paper, there is a high likelihood that no significant differences will be found, so proper range adjustments may not always be performed.

one-factor designs, **randomized block designs**, **nested designs**, **mixed designs**, **hierarchical designs**, and countless other tests? Can we really prepare reviewers capable of reviewing power analysis for these tests?

For reviewing pre-registration content (3), detailed knowledge of multiple comparisons and p-value adjustments is necessary. Why were those specific sub-tests and multiple comparisons pre-registered? Is the range of p-value adjustments for levels, attributes, and criterion variables appropriate as registered? Are the p-value adjustments calculated from the registered preliminary test schedules correct? Reviewers must make these judgments based on evidence. It may be feasible for journals in statistics, but can such reviewers be secured for psychology and substantive sciences that use statistics as a tool? As a result, some academic journals that have implemented the pre-review system have declared, "For the time being, we will not pre-review justifiable n," and such journals indeed already exist.

If we do not pre-review (1), (2), and (3), it will be impossible to prevent the lack of reproducibility caused by the invisible hand of God, the zombie problem, and multiplicity problems. Even with the implementation of the pre-registration system, it becomes ineffective regarding problems caused by significance testing. If significance testing continues to be a standard in peer review, pre-registration/review of (1), (2), and (3) is essential. However, this represents a significant burden for the editors of academic journals. Unfortunately, most pre-registration systems currently implemented in our country are ineffective. Zombies can freely devour the reproducibility of research.

Secondly, there is the problem of what to do with basic experiments, graduation theses, dissertations, and conference presentations. The lack of reproducibility in research results has become a significant problem because academic value has been judged solely by the criterion of p values below 0.05. The pre-review/registration system was devised as an effective countermeasure to this issue. When practicing kendo, we wear protective gear not just for matches but also for practice, due to the inherent risk of being struck by bamboo swords. Similarly, if we perform significance testing or derive academic insights from testing in educational basic experiments, graduation theses, dissertations, or conference presentations, it is necessary to pre-review/register (1), (2), and (3). Omitting this just because it's not as formal as a submission paper is akin to fighting with bamboo swords without protective gear in practice. Whether or not to submit is independent of the mathematical properties (and shortcomings) of the methodology.

Basic experiments are an important subject in the first-year education of psychology. It's an excellent opportunity to learn how to prove that a test is not zombie-like. Missing this opportunity is irresponsible as educators. Instructors should pre-register the deviation from the null hypothesis, significance level, and test power to determine the number of observation subjects n before data collection. This is the correct way to teach significance testing. For graduation theses, dissertations, and conference presentations, the mathematical shortcomings that leave no evidence of violation remain the same. Therefore, if significance testing is performed, instructors must ensure that the tests are pre-registered.

Especially for dissertations, the supervisor must guarantee their academic quality. Supervisors have a professional obligation to prove that the tests are not zombie-like. However, this is more challenging than it seems. Even if the supervisor rigorously seals the pre-registration and shows a timestamp, it can be countered with arguments like "that can be rewritten by software." Recently, there are free websites for pre-registration, which should definitely be used for educational purposes like basic experiments. But this means trusting the website operator. Pre-registration for conferences, too, is essentially the same in that it's based on trust.[9]

The most reliable method of pre-registration at present is using **blockchain**, which is employed for transaction records of cryptocurrencies like Bitcoin. However, even this is not foolproof; if someone with technical skills gets hold of the key in a hot wallet state, rewriting is possible. "Ensuring that the decision was made in advance" is a very difficult task to guarantee permanently.

4.6 Conclusion

The necessity of intent to specify a test is essential knowledge that has not been taught in statistical education. It is hardly mentioned in current textbooks, representing a complete omission in elementary statistical education. However, viewed differently, this omission is perhaps because the nature of the requirement is so unnatural and inconvenient. We need a method that does not depend on the **intent of data collection**. It is natural to monitor research subjects without malice, analyze them several times along the way, and proceed with the research while observing the situation. The theoretical structure of significance testing, which does not allow this, is highly inconvenient. We need a method that allows for **monitoring at any point along the way**.

If we are to use significance testing, pre-determining n based on power analysis is essential, regardless of whether the study is submitted for publication. However, permanently guaranteeing that n was pre-determined before data collection in educational basic experiments, graduation theses, dissertations, and conference presentations is a tremendous effort. Pre-review of power analysis in submitted papers is not a light burden in terms of human resources. We need a method that **does not require pre-determining** n. The deviation from the null hypothesis, as indicated by power analysis, is not always valid as a measure of the effect of research. Power analysis, which cannot appropriately reflect the free research goals arising from real-world demands, is inconvenient.

---

[9] In 2018, the Japanese Ministry of Finance, which was supposed to be trusted by the public, altered official documents related to the Moritomo Gakuen case. The pre-registered content managed by an academic society could also be rewritten due to considerations, etc., in principle.

Interlude: Problems Revealed in Part I

This book is a mystery novel. So far, no villains have appeared. However, numerous complex incidents have occurred one after another, which we will summarize below.

1. "Statistically Significant" Is Merely a Necessary Condition
Rejecting the null hypothesis (e.g., $H_0 : \mu_{\text{difference}} = 0$ or $H_0 : \pi = 0.5$) is merely a very modest necessary condition for academically useful results. This necessary condition is guaranteed at a certain risk level. Being human is a necessary condition for participating in the Olympics, but not everyone who is human can participate. "Statistically significant" as a mere necessary condition is not a valid criterion for academic paper acceptance. It is akin to allowing participation in the Olympics just for being human, which does not ensure empirical reproduction of academically valuable phenomena. It leads to the inclusion of academically meaningless papers in journals.

2. The Invisible Hand of God
In journals where a difference is considered significant if the p-value is less than 5%, the invisible hand of God results in a proliferation of academically meaningless papers. This is not a problem of researchers' ethics or conscience but a natural consequence of contradictions in the peer review system. The very fact that situations like "too much data" arise in data analysis itself signifies an essential flaw in the theoretical structure of significance testing.

3.1. Inability to Actively Demonstrate No Difference
Statistical methodologies capable of substantively demonstrating "no difference" are earnestly requested from the scientific community. For example, showing no difference in the distribution of conditions between experimental and control groups before administering antibiotics A and B actively demonstrates more appropriate control experiments. Hence, even in prestigious journals, the misuse of concluding no difference based on non-significance has been repeated. However, significance testing cannot actively support the null hypothesis. Concluding no difference based on non-significance is like declaring guilt solely for lack of an alibi.

3.2. The Necessity to Determine n Before Data Collection
If n is too large (the wolf), the test becomes overly powerful, and statistically significant but scientifically meaningless reports are made. If n is too small (the tiger), the test power is too low, and potentially significant scientific discoveries are buried without reaching statistical significance. With a tiger at the front gate and a wolf at the back, if you do use significance testing, you cannot just decide n arbitrarily. If you use significance testing, you must include the rationale for determining the number of observation subjects in advance based on power analysis in your paper or report. However, power analysis itself has various shortcomings. Moreover, the policy of determining n in advance is inconvenient and restrictive.

3.3. "Effect Size" in Power Analysis May Not Always Be Valid as a Measure of Effect

The deviation from the null hypothesis used in power analysis is sometimes referred to as the "effect size." However, "effect size" is not always valid as a measure of the research effect. For example, in power analysis for paired differences in means, $\delta_{\text{difference}}$ is used as the "effect size." But the effect size that matches actual experience is often realized after calculating $s_{\text{difference}}$. Moreover, "effect size" is not the only indicator of effect magnitude. Yet, $\delta_{\text{difference}}$ must be selected, which is inconvenient. Indicators for evaluating the research effect should be chosen more freely, prioritizing demands from reality, not constrained by the "effect size" in power analysis.

Column: Multiplicity Problems

Performing multiple significance tests for levels, attributes, criterion variables, periods, or combinations thereof leads to inflation of α. This is known as the multiplicity problem in significance testing. There are countermeasures like multiple comparison corrections for this problem, but they bring their own issues, such as snooping. More serious is the exponential increase in the number of tests due to combinations of levels, attributes, criterion variables, and periods. Correctly adjusting for this often results in no significant differences across the entire report. Therefore, hyperinflation due to combinations is often conveniently ignored.

4. The Zombie Problem

Intent 4 involves increasing observation subjects while repeating significance tests and monitoring p-values. The apparent final output of tests under Intent 4 is indistinguishable from those conducted under textbook Intent 1. However, it ravages the important fruit of research reproducibility. These indistinguishable tests are referred to as zombie-like tests (or zombies), analogous to philosophical zombies. To avoid infection by zombies, power analysis must be conducted, and observation subject numbers must be determined in advance, fixing to "Intent 1 human." It's crucial not to let zombies infected unknowingly lie about being "human" and for uninfected humans to prove they are not zombies. For this, the number of observation subjects determined in advance must be registered with a third party. This third party must permanently guarantee that the registration was made in advance. Eradicating zombies incurs significant costs.

Zombie-like confidence intervals derived solely from the principles of testing inherit the flaws of testing and easily become zombies. The confidence intervals for Intent 3 zombies are not originally calculable. The confidence intervals for Intent 4 zombies become inappropriately narrow. Bem might have conducted tests at a 70% risk rate under the guise of 5%, showing narrow zombie confidence intervals. There is no evidence to refute this concern.

Interlude: Issues Revealed in Part I

This book is a mystery novel. Until now, no villains have appeared. However, a series of challenging cases have occurred, which we summarize below:

1. "Statistically Significant" Is Merely a Necessary Condition
Rejecting the null hypothesis (e.g., $H_0 : \mu_{(\text{difference})} = 0$ or $H_0 : \pi = 0.5$) is merely a very modest necessary condition for academically useful results. It guarantees this necessary condition at a certain risk level. Being human is a necessary condition to participate in the Olympics, but not all humans can participate. Statistically significant as a necessary condition is not a valid criterion for academic paper acceptance. It's like allowing someone to participate in the Olympics just because they are human. This does not guarantee the empirical replication of phenomena with academic value, leading to the inclusion of academically meaningless papers in journals.

2. The Invisible Hand of God
In a journal where a p-value of less than 5% is accepted as a significant difference and the judgment is automatically weighted, the invisible hand of God would be full of academically meaningless papers. This is not an issue of researchers' ethics or conscience but a natural consequence of contradictions in the peer review system. In the first place, the self-negating situation of having too much data in a data analysis is itself an essential flaw in the theoretical construct that significance tests possess.

3.1. Inability to Actively Demonstrate "No Difference"
There is a dire need in science for statistical methodologies that can actively demonstrate "no difference." For instance, if one could show no difference in the condition distribution between experimental and control groups before administering antibiotics A and B, it would actively indicate more appropriately conducted control experiments. Hence, even in prestigious academic journals, the misuse of interpreting non-significance as no difference has been repeated. "However, significance testing does not allow active support of the null hypothesis. Concluding no difference based on non-significance is akin to declaring guilt without an alibi."

3.2. Pre-determined Sample Size n Is Necessary
(wolf) If n is too large, the test becomes overly powerful, leading to statistically significant but scientifically meaningless reports. (tiger) If n is too small, the test lacks power, burying scientifically meaningful discoveries in insignificance. Faced with a tiger at the front gate and a wolf at the back, one cannot casually decide n if opting for significance testing. When conducting a significance test, the paper or report must always include an argument as to why the number of observations was chosen a priori, along with a power analysis. However, even this approach has various drawbacks, and the policy of deciding n in advance is inconvenient and restrictive.

3.3. "Effect Size" in Power Analysis May Not Always Be Appropriate
The deviation from the null hypothesis used in power analysis is sometimes referred to as "effect size." However, "effect size" is not always a valid measure of

the study's effect. For example, in power analysis for paired mean difference tests, $\delta_{(\text{difference})}$ is pre-determined as the "effect size." However, a more relatable effect size, $s_{(\text{difference})}$, often becomes apparent only after analysis. Moreover, "effect size" is not the sole indicator of an effect's magnitude, yet $\delta_{(\text{difference})}$ must be chosen. This is inconvenient. Indicators for evaluating research effects should prioritize demands from reality and not be limited to the "effect size" in power analysis.

Column: Multiplicity Problem

When conducting multiple significance tests for levels, attributes, criterion variables, or time periods, either singly or in combination, α inflation occurs. This is known as the multiplicity problem in significance testing. There is a solution to the multiplicity problem called multiple comparison. However, there are many ways to handle multiple comparisons, and snooping can be a problem. More severely, the combination of levels, attributes, criterion variables, and time periods increases the number of tests exponentially. Correct adjustment leads to no significant differences throughout the report. This often leads to turning a blind eye to hyperinflation caused by combinations.

4. Zombie Problem

Continuously performing significance tests while increasing observations and monitoring p-values is known as intention 4. The final output of intention 4 tests is indistinguishable from those explained in textbooks as intention 1. However, it devastates the crucial fruit of research reproducibility. Such indistinguishable tests are metaphorically termed "zombie tests" (or zombies). To avoid infection, one must determine the number of observations based on power analysis before data collection, fixating on "intention 1 humans." Do not allow infected zombies, who are indistinguishable from each other in appearance, to lie and say that they are "human." It is also important that the uninfected person be able to prove that he or she is a human being and not a zombie. This requires prior registration of the fact that the number of observations has been determined in advance with a third party, who must maintain a permanent guarantee that the registration has been made in advance. The third party must maintain a permanent after-the-fact guarantee that the registration has been made in advance. Eradicating zombies incurs immense costs.

Confidence intervals derived solely from the principles of testing inherit the flaws of testing and easily become zombies. Intention 3 Zombie confidence intervals are inherently unobtainable. Intention 4 zombie confidence intervals become unjustifiably narrow. Perhaps Bem tested at a 70% risk level while pretending it was 5%, showing narrow zombie confidence intervals. There is no evidence to negate this concern.

Part II
Saving the Field of Statistics!

Chapter 5
Bayes' Theorem—"The Probability of a Research Hypothesis Being Correct"

An Example with Estimation of Ratios.

Part II is the solution section. Here, we discuss methods to overcome the problems presented in Part I.

5.1 The Probability of Being Spam

In our modern society, we rely on Internet emails daily. While emails are incredibly convenient, those filled with advertisements, solicitations, or scams are a significant nuisance. A plethora of spam emails can bury important messages. To prevent this, spam filters are used. These filters calculate the **probability of an email being spam** for each message that arrives at the mail server. Emails exceeding a certain probability threshold are not placed in the main mailbox but are instead quarantined in a dedicated box. Thanks to these filters, we, the users, are spared from viewing spam emails. Here, we will learn how to calculate the probability of an email being spam.

5.1.1 Joint Probability and Conditional Probability

Imagine a scenario where an email server receives 100 emails as shown in Table 5.1. Consider an experiment where one email is randomly selected.

H. Toyoda, *Statistical Significance and the PHC Curve*,
https://doi.org/10.1007/978-981-97-7748-8_5

Table 5.1 Breakdown of 100 emails in the box

	With word B	Without word ¬B	Total
Spam email A	37	11	48
Regular email ¬A	19	33	52
Total	56	44	100

We represent the probability of observing a certain event with $f()$. For example, the probability of observing the event A, which is an email being spam, is 48 out of 100, so

$$f(\text{A}) = 0.48 \tag{5.1}$$

The symbol ¬ represents negation. For instance, ¬A means the email is not spam, a regular email, and $f(\neg\text{A}) = 0.52$. The sum of $f(\text{A})$ and $f(\neg\text{A})$ is always 1. Let's define the event of observing a specific word B in an email as B. Word B could be, for example, "investment guide."

$$f(\text{B}) = 0.56, \quad f(\neg\text{B}) = 0.44 \tag{5.2}$$

Again, their sum is 1.

The probability of two events occurring simultaneously is known as **joint probability**. For instance, the joint probability of having word B in an email and that email being spam is

$$f(\text{A}, \text{B}) = f(\text{B}, \text{A}) = 0.37 \tag{5.3}$$

This represents the probability of an email being spam and containing word B. The equality sign connects the left and middle terms.

Now, let's consider the probability of containing word B in an email given that it is known to be spam. The probability of one event occurring under the condition that another event has been observed is called **conditional probability**. Out of 48 spam emails, 37 meet this condition, so the probability is

$$f(\text{B}|\text{A}) = 37/48 \simeq 0.77 \tag{5.4}$$

The vertical bar | read as "given" as introduced in Chap. 1. Equation (5.4) expresses the probability of event B occurring given that event A has been observed. Equation (1.4) used the format (data|parameter). The vertical bar | represents a conditioning relationship that does not distinguish between events, data, or parameters and will be frequently used hereafter.

Conversely, let's consider the probability that an email is spam given that it contains word B. This probability is

$$f(\text{A}|\text{B}) = 37/56 \simeq 0.66 \tag{5.5}$$

As shown in Eq. (5.3), the value of joint probability does not change with the order of events. However, as shown in Eqs. (5.4) and (5.5), the value of conditional probability does change depending on the order of events on either side of the vertical bar.

5.1.2 Bayes' Theorem

There exists a relationship between joint probability and conditional probability as follows:

$$f(\mathrm{A}, \mathrm{B}) = f(\mathrm{B}|\mathrm{A})f(\mathrm{A}) \tag{5.6}$$

By inserting actual values, we get:

$$0.37 = (37/48) \times 0.48 \tag{5.7}$$

which verifies the validity of Eq. (5.6). Joint probability and conditional probability can be expressed by interchanging events A and B, and using Eq. (5.3) as:

$$f(\mathrm{A}, \mathrm{B}) = f(\mathrm{A}|\mathrm{B})f(\mathrm{B}) \tag{5.8}$$

Similarly, we can confirm:

$$0.37 = (37/56) \times 0.56 \tag{5.9}$$

verifying the validity of Eq. (5.8).

Since Eqs. (5.6) and (5.8) have a common left-hand side, we can connect the right-hand sides:

$$f(\mathrm{A}|\mathrm{B})f(\mathrm{B}) = f(\mathrm{B}|\mathrm{A})f(\mathrm{A}) \tag{5.10}$$

Dividing both sides by $f(\mathrm{B})$ gives:

$$f(\mathrm{A}|\mathrm{B}) = \frac{f(\mathrm{B}|\mathrm{A})f(\mathrm{A})}{f(\mathrm{B})} \tag{5.11}$$

This equation is known as **Bayes' Theorem**, or **Bayesian Formula**. By substituting numbers, we find:

$$0.66 \simeq (37/56) = \frac{(37/48) \times (0.48)}{0.56} \tag{5.12}$$

confirming the validity of Bayes' Theorem.

The event whose probability of occurrence we want to investigate is event A, the email being spam. In Bayes' Theorem, the numerator $f(\mathrm{B}|\mathrm{A})$ on the right-hand side represents information from the data related to the event of interest. Similarly, $f(\mathrm{A})$

in the numerator is the probability of event A before using information from the data, known as the **prior probability**. The left-hand side $f(\mathrm{A}|\mathrm{B})$ is the probability of event A occurring after using information from the data, known as the **posterior probability**.

The information from the data was $f(\mathrm{B}|\mathrm{A}) \simeq 0.77$. With the addition of the information "containing suspicious word B," the posterior probability of the email being spam (0.66) is higher than the prior probability (0.48).

5.1.3 *Probability of Not Being Spam*

The prior probability of an email being spam is 0.48, and the prior probability of not being spam, $f(\neg\mathrm{A})$, is 0.52, adding up to 1. Then, what is the probability of an email not being spam if word B is observed? The posterior probability is:

$$f(\neg\mathrm{A}|\mathrm{B}) = \frac{f(\mathrm{B}|\neg\mathrm{A})f(\neg\mathrm{A})}{f(\mathrm{B})} \tag{5.13}$$

$$0.34 \simeq (19/56) = \frac{(19/52) \times 0.52}{0.56} \tag{5.14}$$

Regarding posterior probabilities, the sum of the probability of being spam (0.66) and not being spam (0.34) adds up to 1.[1] Bayes' Theorem can be easily understood even by middle school students.

5.1.4 *Transformation of Bayes' Theorem 1*

Equations (5.11) and (5.13) share the common denominator on the right-hand side, $f(\mathrm{B})$, which does not contain information related to the event of interest, namely whether the email is spam or not. Therefore, the probability of the email being spam or not can be expressed using a proportional relationship:

$$37 : 19 = f(\mathrm{A}|\mathrm{B}) : f(\neg\mathrm{A}|\mathrm{B}) = f(\mathrm{B}|\mathrm{A})f(\mathrm{A}) : f(\mathrm{B}|\neg\mathrm{A})f(\neg\mathrm{A})$$

Using the condition that their sum is 1, this proportional relationship can be uniquely determined as $0.66 : 0.34$ without $f(\mathrm{B})$. Therefore, Bayes' Theorem is often handled in the form:

[1] Throughout this book, we will calculate the probability of numerous research hypotheses being correct. In every case, without exception, the sum of the probabilities of the hypothesis being correct and not correct adds up to 1. In significance testing, the sum of the two competing errors is $\alpha + \beta \neq 1$, which has been a detriment to elementary education. The Bayesian method is clear in this regard.

$$f(\text{A}|\text{B}) \propto f(\text{B}|\text{A})f(\text{A}) \tag{5.15}$$

where $\propto$ denotes a proportional relationship between both sides.

5.1.5 Transformation of Bayes' Theorem 2

Now, let's focus on the denominator $f(\text{B})$ in Bayes' Theorem that was omitted in the previous section. The probability $f(\text{B})$ of word B appearing in all emails is the sum of the probabilities $f(\text{B}, \text{A})$ for spam emails and $f(\text{B}, \neg\text{A})$ for non-spam emails. Further, using Eq. (5.6):

$$f(\text{B}) = f(\text{B}, \text{A}) + f(\text{B}, \neg\text{A}) = f(\text{B}|\text{A})f(\text{A}) + f(\text{B}|\neg\text{A})f(\neg\text{A}) \tag{5.16}$$

Substituting the values from Table 5.1, we confirm:

$$0.56 = 0.37 + 0.19 = (37/48) \times 0.48 + (19/52) \times 0.52 \tag{5.17}$$

This expression can be substituted in the denominator of Eq. (5.11) to obtain:

$$f(\text{A}|\text{B}) = \frac{f(\text{B}|\text{A})f(\text{A})}{f(\text{B}|\text{A})f(A) + f(\text{B}|\neg\text{A})f(\neg\text{A})} \tag{5.18}$$

This equation is called a transformed form of Bayes' Formula.

If word B is not in the email, the posterior probability of the email being spam, with B replaced by $\neg$B, is determined by:

$$f(\text{A}|\neg\text{B}) = \frac{f(\neg\text{B}|\text{A})f(\text{A})}{f(\neg\text{B}|\text{A})f(A) + f(\neg\text{B}|\neg\text{A})f(\neg\text{A})} \tag{5.19}$$

Using numbers from Table 5.1, we get:

$$0.25 = (11/44) = \frac{(11/48) \times 0.48}{(11/48) \times 0.48 + (33/52) \times 0.52} \tag{5.20}$$

Since suspicious word B was not observed, the posterior probability of the email being spam (0.25) is less than the prior probability (0.48).

5.1.6 Bayesian Updating

The probability of an email being spam is not calculated from a single piece of data. For example, upon examining the appearance probabilities of words C, D, and E in a

Table 5.2 Appearance probabilities of words B, C, D, E

	Investment guide B	Completely free C	Sure surge D	Winning investment E
$f(\mid A)$	(37/48)	0.85	0.83	0.88
$f(\mid \neg A)$	(19/52)	0.11	0.09	0.08

large volume of daily emails, we find the data as shown in Table 5.2. For instance, the second row in the third column indicates that the probability $f(C|A)$ of the phrase "completely free" appearing in spam emails is 0.85. The third row in the same column shows that the probability $f(C|\neg A)$ of it appearing in regular emails is 0.11.

Bayes' theorem can be used sequentially. We reinterpret the posterior probability calculated so far as the prior probability, and with each new piece of data, we calculate a new posterior probability. This is known as **Bayesian updating**. **Today's posterior probability becomes tomorrow's prior probability**.

As we read a particular email, not only do we find word B, but also words C, D, and E appear one after another. It is an email full of suspicious words. Appropriately modifying Bayes' theorem (Eq. 5.18) and applying it sequentially three times, we calculate the probability that the email is spam:

$$0.9375 = \frac{0.85 \times 0.66}{(0.85 \times 0.66) + 0.11 \times (1 - 0.66)}$$

$$0.992823 = \frac{0.83 \times 0.9375}{(0.83 \times 0.9375) + 0.09 \times (1 - 0.9375)}$$

$$0.9993433 = \frac{0.88 \times 0.992823}{(0.88 \times 0.992823) + 0.08 \times (1 - 0.992823)}$$

From an almost fifty-fifty probability of 48%, it rapidly increases to 66, 94, 99, and 99.9%, approaching 1.0 for the probability of being spam (and 0.0 for being a regular email).[2] It's akin to the natural progression of a person's concerns. This demonstrates how Bayes' theorem can be applied sequentially to each piece of data. While we omit the details in this example, it's also possible to apply Bayes' theorem all at once to data concerning words B, C, D, and E. The method of simultaneous application will be studied in the following sections through specific examples.

If we continue reading the email and find that word B is not present, as in the situation of Eq. (5.20), and furthermore, words C, D, and E are also not found, meaning no suspicious words are present. Appropriately modifying Eq. (5.19) and applying it sequentially three times, we calculate the probability that the email is spam:

[2] The left-hand side of the equation takes many significant figures to avoid cumulative computational errors.

$$0.053 = \frac{(1-0.85)\times 0.25}{((1-0.85)\times 0.25)+((1-0.11)\times(1-0.25))}$$
$$0.010386 = \frac{(1-0.83)\times 0.053}{((1-0.83)\times 0.053)+((1-0.09)\times(1-0.053))}$$
$$0.001367 = \frac{(1-0.88)\times 0.010386}{((1-0.88)\times 0.010386)+((1-0.08)\times(1-0.010386))}$$

The probability of being spam plummets rapidly from an almost fifty-fifty chance of 48% to 25, 5, 1%, and finally to 0.1%, approaching 0.0 (and 1.0 for being a regular email).

5.1.7 Comparison with p-Values

The probability of being spam used in spam filters has markedly superior properties compared to p-values.

Property 1 It is an index that can be understood by those with domain knowledge. A statement like "There is a 99% probability of this being spam" is straightforward and understandable even to elementary school students. It is not about verifying necessary conditions. It directly shows the probability of the proposition of interest being true. If such a tangible indicator were the conclusion of a research paper, Bem's paper would not have been published. The p-value in Eq. (1.10) does not allow for an academic interpretation in the respective field and is unrelated to tangible insights.

Property 2 As the number of observations increases, the determination of the truthfulness of the hypothesis becomes clearer. The more data added, the clearer the truth or falsity of the hypothesis becomes. There can be true or false outcomes, bringing a definitive resolution. It seems possible to escape from the invisible hand of God.

There can never be too much data. This is a desirable property in the era of big data. In contrast, with significance testing, an increase in data can lead to meaningless significant differences.

Property 3 Unlike significance testing, no matter how many times calculations are performed along the way, it does not adversely affect subsequent calculations. Data can be freely added while monitoring the situation. In this sense, it is like daily weight measurements. Today's analysis results are not negated by yesterday's analysis. It seems like a potential remedy against the type 4 zombie problem.

This property is particularly noteworthy. It eliminates the need for power analysis and the necessity to decide on n in advance and to continuously ensure that this decision was made in advance. In this sense, the need for prior registration[3] is obviated.

[3] The purpose of prior registration is not only to correct distortions in p-values. This issue is discussed in Chap. 8. However, using Bayesian methodology, at least the need to decide on n in advance is eliminated.

5.2 Bayes' Theorem for Distributions

Bayes' theorem is a very universal theorem, allowing substitution of events, data, and parameters. Replacing A and B in Eq. (5.11) with symbols θ and x, respectively, we get:

$$f(\theta|x) = \frac{f(x|\theta)f(\theta)}{f(x)} \tag{5.21}$$

Here, θ represents a general notation for parameters. **Parameters** are numerical indicators that determine the characteristics of a probability distribution. For example, population proportion π, population mean μ, and population standard deviation σ are parameters, and the symbol θ is used to represent these collectively. x represents data, such as the number of successes, length of hospital stay, or a set of weight measurements.

The left side of Eq. (5.21) represents the distribution of the parameter θ given the data x, known as the **posterior distribution**. In Eq. (5.11), we discussed the probability of an email being spam, hence called it the posterior probability. But here, we are showing the likely distribution of population parameters like the population proportion or mean given the data, hence we call it the posterior distribution. The numerator $f(\theta)$ on the right side of Eq. (5.21) is the distribution of the parameter θ before the data x is given, called the **prior distribution**.

The denominator $f(x)$ on the right side of Eq. (5.21) is known as the **normalizing constant** and represents the probability of observing the data. However, the normalizing constant $f(x)$ does not include the parameter of interest θ. Therefore, considering Eq. (5.15), Eq. (5.21) is modified to:

$$f(\theta|x) \propto f(x|\theta)f(\theta) \tag{5.22}$$

In this book, Bayes' theorem is consistently treated in the form of Eq. (5.22). When using the Markov Chain Monte Carlo (MCMC) method,[4] the normalizing constant $f(x)$ can remain unknown, yet the posterior distribution on the left side can still be determined.

5.3 Likelihood Principle

The right side of Eq. (5.22), $f(x|\theta)$, is the probability of observing the data given the parameter. When interested in the population proportion, $f(x|\theta)$ is denoted as $f(x|\pi)$. Recall the example of the treatment method introduced in Sect. 4.2. There,

[4] Those interested in the theoretical foundation of the Markov Chain Monte Carlo method, please refer to Chaps. 4 and 5 of [15].

out of 24 treated individuals, 7 did not recover. Let's denote the cured as "Cure" (C) and not cured as "Non-cure" (Nc), and suppose the treatment results for 24 people in the order they were determined are:

$$\begin{aligned} x = (&\text{C, C, Nc, C, Nc, C, Nc, C, C, C, Nc, C,} \\ &\text{C, Nc, C, C, Nc, C, C, C, C, C, C, Nc)} \end{aligned} \tag{5.23}$$

The sample distribution derived from this data varies depending on the unseen intentions of how the data was observed. Naturally, the p-values also change. This is a fatal property of significance testing.

5.3.1 Likelihood

The probability of observing this binary data in this particular order is expressed as:

$$f(x|\pi) = \pi^7(1-\pi)^{24-7} \tag{5.24}$$

The right-hand side is called the kernel (essential part) of the binomial distribution and shares similarities with the kernel of the negative binomial distribution. For instance, the probability of getting heads with a fair coin (population proportion $\pi = 0.5$) twice in two tosses is 0.25. Probability refers to terms used when the parameter is fixed and the data are treated as variables. Conversely, in this example, the parameter is unknown, and we aim to estimate it. Here, the observable data is fixed, and the parameter is treated as a variable. The roles of constants and variables are switched. In such cases, to avoid confusion, the same probability formula is renamed as **likelihood**. Going forward, $f(x|\pi)$ or its general form $f(x|\theta)$ will be referred to as likelihood.

5.3.2 Likelihood Does Not Depend on Intent

Likelihood is independent of intents 1 through 4. Whether a researcher starts a clinical trial with the intention of:

(1) Treating 24 patients and ending up with 7 non-cures, or
(2) Testing for treatment effect after failing to cure 7 patients and the 24th patient being the 7th non-cure, or
(3) Setting a research period of one year and testing the treatment outcomes of patients who visit during that period, resulting in 24 patients treated with 7 non-cures, or
(4) Increasing the observation count while performing binomial tests for each case, monitoring the p-values, and deciding to only publish the results of the final bino-

mial test if significant difference is observed, even if the interim state involved treating 24 patients, the probability of observing the binary data in the sequence of (5.23) given the parameter is always expressed by (5.24). There is no need to distinguish between them. This is a notable advantage of likelihood. Therefore, there is a viewpoint that statistical inference should be based on likelihood, known as the **Likelihood Principle** [16].

The theory of significance testing indispensably includes the concept of a **sample distribution**. The sample distribution was an imaginary distribution of a statistic that was not calculated in repeated experiments under a specific intent. Mathematical simple intent 1 hardly exists in natural research settings, and most often the real intent was resource-fixed intent 3, which does not lead to a sample distribution. Pretending to be textbook intent 1 while actually performing p-value monitoring intent 4 caused an unacceptable inflation of α. Moreover, after the fact, it's extremely unlikely to have evidence that it was actually intent 4. Due to the significant adverse effects and the unlikelihood of leaving evidence, the system of prior registration was reluctantly developed.

Inference based on likelihood does not require the concept of a sample distribution. Regardless of the computation schedule, whether monitored or not, even if data collection is stopped midway or additional data is collected, the likelihood remains unchanged. There is no need for power analysis and pre-determining n, hence no need for prior registration (1) (2) (3) essential for significance testing in Sect. 4.5. One can focus solely on the data actually observed.

By not using sample distributions, one is freed from one of the shortcomings of significance testing: the multiplicity issue. Of course, this also eliminates the need for multiple comparisons, one aspect of the multiplicity problem.

5.3.3 Prior and Posterior Distributions

To estimate the population proportion, we replace θ with π in Bayes' theorem as expressed in Eq. (5.22), resulting in:

$$f(\pi|x) \propto f(x|\pi)f(\pi) \tag{5.25}$$

The prior distribution $f(\pi)$ represents the distribution of the parameter before observing any data. In formal analyses like in papers and reports, it should not be narrowly concentrated on a specific area without reason. A prior distribution that does not concentrate on a specific area is called a **non-informative prior**.

The population proportion π is defined within the interval [0, 1]. Therefore, the prior distribution of π, $f(\pi)$, before observing data, is considered a uniform distribution over the interval [0, 1]. Choosing a uniform distribution as the prior means that probabilities related to where the population proportion might lie within its domain are equal. This is why it is considered non-informative.

In the interval [0, 1], a uniform distribution makes $f(\pi)$ constant. This equivalence to a state without including π leads to the posterior distribution in Eq. (5.25), which is proportional only to the likelihood:

$$f(\pi|x) \propto f(x|\pi) \tag{5.26}$$
$$= \pi^x(1-\pi)^{n-x} \tag{5.27}$$

The posterior distribution is proportional only to the likelihood (or to the kernel of the binomial distribution), leaving no trace of the prior distribution. This allows reflecting only the information from the likelihood (data) in the posterior distribution. This unique advantage is given by the uniform distribution, making it non-informative in this sense too.

Once the posterior distribution is determined, the MCMC method is used to generate random numbers $\pi^{(1)}, \pi^{(2)}, \ldots, \pi^{(t)}, \ldots, \pi^{(T)}$ to approximate the posterior distribution of the population proportion π.

5.4 Estimating the Population Proportion for "Premonition of the Future"

Let's estimate the population proportion for Bem's "Premonition of the Future" mentioned in Chap. 1. By substituting $x = 829$, $n = 1560$ into Eq. (5.27) and setting $T = 100{,}000$, we get an approximation of the posterior distribution of π as shown in Fig. 5.1.

The summary statistics of the posterior distribution are presented in Table 5.3. The most frequently used point estimate for the parameter is the mean of the posterior distribution, known as the **posterior expected value**. The estimated posterior expected value, calculated as the mean of the random numbers, is referred to as the **EAP (Expected A Posteriori) estimate**. The point estimate is 0.531, which, calculated in the Bayesian way, matches Bem (2011).

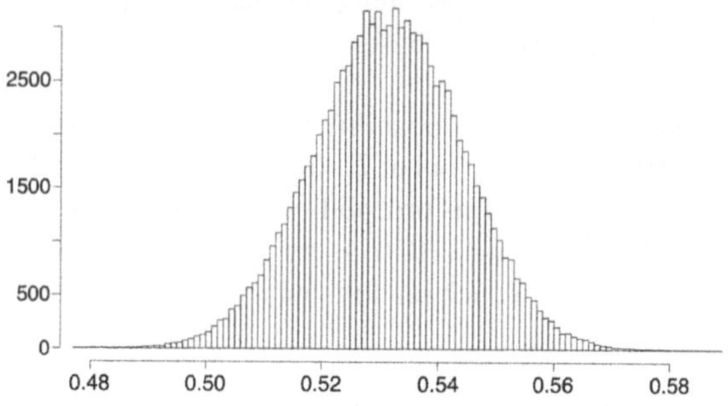

Fig. 5.1 Posterior distribution of the population proportion for "Premonition of the Future"

Table 5.3 Summary statistics of the posterior distribution for "Premonition of the Future"

	EAP	post.sd	2.5%	5%	50%	95%	97.5%
Ratio	0.531	0.013	0.506	0.510	0.531	0.552	0.556

The stability indicator of the posterior expected value is the **posterior standard deviation (post.sd)**, calculated as the standard deviation of the random numbers. The posterior standard deviation was 0.013.

For interval estimation of the parameter, the **95% (two-sided) credible interval**, the interval between the 2.5 and 97.5% points of the posterior distribution, is frequently used. The 95% credible interval was [0.506, 0.556], not including 0.5, thus deviating from the chance level.

In future, when summarizing the posterior distribution, it may be expressed in the format EAP estimate (post.sd) [95% credible interval] (for example, 0.531 (0.013) [0.506, 0.556]) or in the format EAP estimate [95% credible interval] (for example, 0.531 [0.506, 0.556]).

5.4.1 Probability that a Research Hypothesis Is True

The probability that a randomly selected email being "spam" can be directly applied using Bayes' theorem since it is a discrete probability event. However, for statements about continuous quantities, such as parameters (for example, π here), an additional step is necessary to handle the probability of the statement being true. By examining the ratio of occurrences where the research hypothesis holds true (is valid) within the random numbers generated to approximate the posterior distribution, we can determine the **Probability that the Research Hypothesis is Correct** (PHC, or **probability that the hypothesis is correct**) as proposed in literature [8]. In the system, PHC is denoted in uppercase, while the actual probability is denoted in lowercase PHC. Specifically, for a research hypothesis U, we define a binary variable

$$u^{(t)} = \begin{cases} 1 & \text{if research hypothesis } U \text{ is true for } \theta^{(t)},\ (t = 1, \ldots, T) \\ 0 & \text{otherwise} \end{cases} \tag{5.28}$$

and determine the probability that the research hypothesis is correct by examining what fraction of $u^{(t)}$ out of T is assigned the value 1. Simply computing the average of $u^{(t)}$ is sufficient. This probability is denoted as PHC (for the research hypothesis U).

Table 5.4 shows PHC($c < \pi$), varying c from 0.50 to 0.58 in increments of 0.01. Specifically, for each random number, we determine

$$u^{(t)}_{c<\pi} = \begin{cases} 1 & \text{if } c < \pi^{(t)},\ (t = 1, \ldots, 100{,}000) \\ 0 & \text{otherwise} \end{cases} \tag{5.29}$$

Table 5.4 Probability that the research hypothesis 'population proportion greater than baseline c' is true

c	0.50	0.51	0.52	0.53	0.54	0.55	0.56	0.57	0.58
$c - 0.5$	0.00	0.01	0.02	0.03	0.04	0.05	0.06	0.07	0.08
$\mathrm{PHC}(c < \pi)$	0.993	0.953	0.814	0.542	0.248	0.073	0.011	0.001	0.000

and calculate the average of $u^{(t)}_{c<\pi}$. The constant c needed for the calculation of PHC is referred to as the **baseline**. The results, in brief, are as follows:

$\mathrm{PHC}(0.51 < \pi) = 95.3\%$ (Psychic ability once in every 100 trials?)
$\mathrm{PHC}(0.53 < \pi) = 54.2\%$ (At this point, one can no longer make a confident claim)
$\mathrm{PHC}(0.58 < \pi) = 0.0\%$ (Far from the targeted 0.7).

In comparison to p-values, PHC (for the research hypothesis U) provides an intuitive understanding like this and does not require any correction for multiple comparisons like in multiple testing.

5.4.2 *Peer Review with Metrics Convincing to Those with Domain Knowledge*

In significance testing, by rejecting the null hypothesis $H_0 : \pi = 0.5$, peer review has been passed regardless of the research context, analysis objective, applicability, or significance of results across all research fields.

According to Table 5.4, the "probability that the research hypothesis 'population proportion greater than baseline c' is correct" is 99.3% for any value above 0.0, which seems almost certain. However, a 0-point increase does not qualify as "psychic ability." With an increase in the number of observations, even using the Bayesian approach, this probability approaches 1.0, falling into the trap of the "invisible hand." Even in fundamental research, it is wrong to be satisfied with a "greater than 0-point" confidence without a clear reason, and such a review style should be rigorously corrected.

In significance testing, despite referencing deviation from the null hypothesis ($\pi = c$) in power analysis, the test ultimately uses $H_0 : \pi = 0.5$. This approach is half-hearted and counterproductive. Instead, the research hypothesis "existence of psychic ability" should be rephrased as "$c < \pi$, where c is clearly greater than 0.5," and the probability that this probabilistic proposition holds should be considered. In short, peer review must be conducted with metrics that are convincing to those with domain knowledge.

5.4.3 Determining the Baseline c

The term "clearly larger" is ambiguous. How should the baseline c be determined? There are broadly two approaches.

One approach is to examine the distribution of opinions among multiple experts in the relevant field. However, in this case, there are no experts who can assess the value of predicting the location of a nude picture. Therefore, I surveyed students with the question, "If there were precognitive abilities, how accurately could they predict the location of nudity?" Almost all responded, "Since it's a superpower, it should be 100."

Another direction is to evaluate c from the perspective of the specific techniques in the field. I have experience using artificial intelligence for investing about 20 years ago, a conservative mainstay in future prediction. I created a system [6] that automatically collected stock prices from the internet and used a backpropagation-trained neural network to predict the next moment's stock price movements.

Of course, merely rejecting $H_0 : \pi = 0.5$ is not enough. The baseline c for the threshold of "being able to live off day trading" is determined by the wealth ensuring a sufficient number of trials to bring the sample hit rate close to the population hit rate and as a function of fees paid to the securities company. Back then, if the prediction system consistently achieved $\pi = 0.53$, it was sufficient for a living. From this perspective, PHC$(0.53 < \pi) = 54.2\%$ is quite promising. If this were not about predicting nudity locations but a new principle for stock price prediction, it might have been a standard worthy of peer review 20 years ago.

The role of statistics is to provide objective materials for judgment, and at the same time, that's as far as its role extends. The main actor in making judgments should be experts with domain knowledge. PHC is a concrete probability that can be intuitively interpreted. Therefore, experts who are non-statisticians can make judgments using their domain knowledge.[5] The same $\pi = 0.53$ can have entirely different academic values depending on the context (e.g., predicting nudity locations or investing). A statistician without domain knowledge cannot, should not, and likely does not want to make such distinctions.

5.4.4 It Is Not Necessary to Set a Baseline c/PHC Curve

Statistically, the baseline c cannot be determined. As mentioned before, if it can be determined through survey or domain knowledge, that would be ideal. However, in many cases, it is difficult to set a single baseline point.

There is no need to set a baseline. Instead, show a table like in Table 5.4 or a graph like in Fig. 5.2. Figure 5.2 is a graph with the baseline c on the horizontal axis and PHC on the vertical axis, visualizing the content of Table 5.4. We'll call this the

[5] Not just p-values, but also information criteria and Bayes factors are abstract enough to make domain knowledge-based judgments difficult.

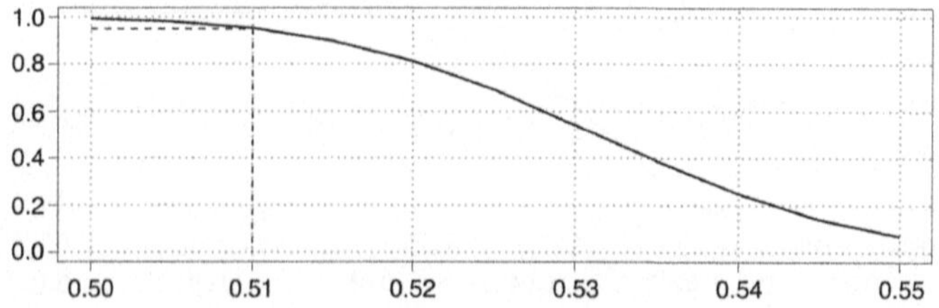

Fig. 5.2 PHC curve (horizontal axis is the baseline c, vertical axis is PHC)

PHC curve. For reference, I have added a dashed line at the baseline where a 95% credible claim can be made.

In cases like this example, where a larger value of the parameter is desirable, we should check PHC($c < \pi$), so the PHC curve will be a monotonically decreasing function.[6] The PHC curve aggregates and represents the research value related to the baseline on the horizontal axis.

Based on the results of the PHC curve, reviewers should focus on the substantive scientific (medical/psychological) implications when deciding whether to accept or reject a given paper. The research value of the same baseline c varies entirely depending on the technical context, so it is no longer a matter for statistics to intervene in.

5.4.5 *Actively Presenting Claims Equivalent to the Null Hypothesis*

Using PHC, not only can we express the probability that "the population proportion is clearly greater than 0.5," but we can also represent the probability of a claim essentially equivalent to the null hypothesis, such as "the population proportion is almost 0.5." However, this is different from the null hypothesis in significance testing and is not represented as PHC($\pi = 0.5$).

Since it's not a mathematical assumption, no one strictly imagines $\pi = 0.5$ for a fair coin toss.[7] The realistic image of a fair coin toss in the real world is $(|\pi - 0.5| < c)$.[8] This corresponds to the hypothesis that "there is no substantial precognitive ability," equivalent to the null hypothesis in significance testing.

For instance, setting $c = 0.05$, we get PHC($0.45 < \pi < 0.55$) $= 0.930$. Under the premise of $c = 0.05$, the probability that the research hypothesis "there is no precognitive ability" is correct is 93.

[6] In this case, it becomes the PHC curve of (1 − cumulative distribution of the posterior distribution).

[7] "Players don't often argue that a coin toss to decide the end and kick-off in soccer is unfair because the sides of the coin are different, making $\pi \neq 0.5$."

[8] This is named **region of practical equivalence** (ROPE, pronounced as "rope") by Kruschke in [14].

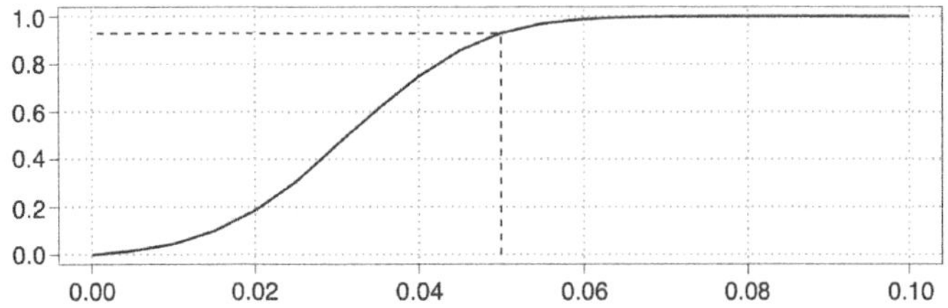

Fig. 5.3 PHC($|\pi - 0.5| < c$) curve

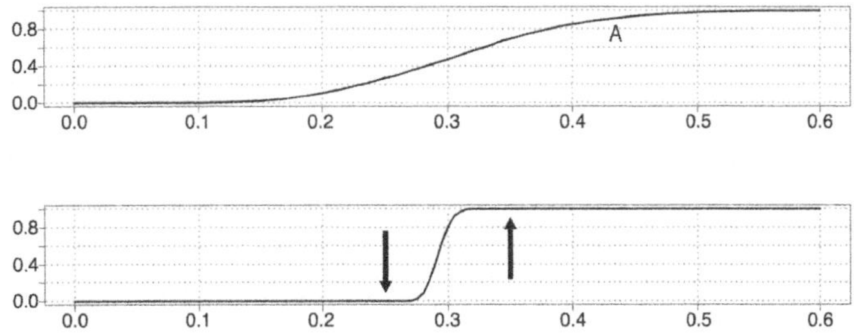

Fig. 5.4 PHC curve for small n (upper graph), large n (lower graph)

To critics (reviewers, class instructors, thesis advisors) who might say, "That ($c = 0.05$) is just the author's subjective opinion," I present a PHC curve corresponding to Fig. 5.3. Figure 5.3 is a PHC curve with the baseline c on the horizontal axis and PHC($|\pi - 0.5| < c$) on the vertical axis. A dashed line is added at the baseline for making a claim with 95% credibility for reference. Observing this curve to judge whether there is precognitive ability or not is a matter of social psychology, not something to be handled by statistics. Even if psychological judgments by reviewers differ, they are each valid and healthy disagreements.

Note that the left end of Fig. 5.3 starts at 0.0. This means PHC($H_0 : \pi = 0.5$) $=$ 0.0, highlighting the essence that the null hypothesis is "strictly false before data collection."

5.4.6 *Clarification of Hypothesis Truth as the Number of Observations Increases*

Using PHC, two benefits arise concerning the number of observations n.

First, there is no irrational problem such as having too much data. The upper graph in Fig. 5.4 is the PHC curve for ($\pi < c$) with $n = 24$, $x = 7$ from Sect. 4.1. In cases like this example, where a smaller value is desirable for the parameter, the PHC curve for PHC($\pi < c$) will be a monotonically increasing function.

The lower graph is the PHC curve for $n = 2400$, $x = 700$. Phc curves using posterior distributions have the general property of changing gently when n is small and steeply as it gets larger. To have high credibility in PHC (non-cure rate is less than the baseline c), only large c in the right region A can be chosen when n is small.[9]

The responsibility for proving research significance lies with the paper's author. In analyses with little data, this property, where favorable claims can only be modestly asserted, is natural and very important. It prevents social disadvantages where results are unstable due to insufficient data, leading to conclusions favorable to researchers by chance. The tiger at the front gate remains necessary for this reason.

PHC (non-cure rate is less than the baseline c) approaches 0 or 1 probabilistically with the increase in n, regardless of the value of c. In the lower graph, PHC($\pi < 0.25$) $\simeq 0$, and PHC($\pi < 0.35$) $\simeq 1$. With many observations, as in the lower graph, results stabilize and clear conclusions can be drawn regardless of whether they are advantageous to the analyst or not. Therefore, there is never a situation of having too much data. The wolf at the back gate is no longer a concern. This also frees us from the invisible hand, a drawback of significance testing.

Second, there is no need to pre-determine the number of observations n. Since we don't use the concept of the sampling distribution, we are free from the issue of multiplicity. Thus, depending on the research budget and resources, data can be incrementally increased until the PHC curve shows the steepness required by the research. There is no need for correction even if PHC is calculated every morning.[10] However, PHC for an effect greater than 0 is not enough. Even using PHC, one can fall into the trap of the invisible hand. It's crucial to use a practically meaningful non-cure rate baseline c and observe a high PHC($\pi < c$).[11] Then, you can start writing your paper the day such a high PHC is observed. In this sense, there is no waste of research resources.

5.4.7 The Binary Judgment of Significance or Non-significance Involves Great Social Disadvantages

For fundamental research that could liberate humanity from energy problems, such as the "discovery of the principle of cold fusion in a beaker," there is a need for swift publication. Such work should be accepted solely on the basis of verifying necessary conditions (rejecting the null hypothesis) if it convinces reviewers and states the necessity of rapid publication. However, this is rather a rare exception. The current peer review system, under the excuse of "it's fundamental research," often defaults to

---

[9] As evident from Fig. 5.2, to have high credibility that the parameter is greater than the baseline c, only small c in the left disadvantageous region can be chosen when n is small.

[10] Because the principle is the same as sequentially calculating "the probability of being spam."

[11] No matter how much data is collected and how many times PHC is calculated, Bem would likely not have observed high credible PHC($0.8 < \pi$). Meaningless papers, even if significant, cannot be published anymore.

merely confirming necessary conditions, leading to significant social disadvantages due to the invisible hand and zombie problems. The desire to know "whether there is an effect or not" is a self-serving bias. For instance, the hypothesis in fundamental research that "symmetrical faces are more attractive than asymmetrical ones" would still be false under the null hypothesis even if 50 million people opposed it and 50,000,001 supported it. Fundamental research cannot be unconditionally accepted just because it confirms necessary conditions. The process of always discussing the extent of effects and drawing PHC curves should become the norm in peer review.

5.5 Conclusion

Using PHC offers the following advantages:

1. If we avoid using PHC for the claim of no effect at all, we are freed from the invisible hand, and papers that are significant but meaningless cannot be published.
2. By calculating the PHC for ROPE, we can actively assess the validity of hypotheses equivalent to the null hypothesis.
3. Attention must be paid to the tiger at the front gate (the problem of having too little data) to avoid self-harm and to make clear claims. However, there is no need to worry about the wolf at the back gate (the problem of having too much data).
4. There is no need to pre-determine the number of observations n; analysis results can be monitored at any stage, and there is no need to distinguish between intentions 1, 2, 3, and 4. Thus, there is no longer a need for pre-registration of n.
5. The PHC curve aggregates and demonstrates the value of the data for the indicators of interest. Therefore, analysts do not need to set a specific baseline c before or after data collection.
6. The PHC indicating "no precognitive ability" ($\pi < 0.7$) is a probability based on the mathematical premise of choosing a binomial distribution for the likelihood and a uniform distribution for the prior. This premise is no different from p-values and various statistical metrics, and PHC is simply the probability that a research hypothesis is correct, with mental brackets added. Keeping this in mind, PHC provides insights that are far more tangible than p-values, which will be detailed in the next chapter.

Chapter 6
Concluding Remarks with Sincerity

An Example with the Inference of Differences between Two Independent Groups.

The p-value of a significance test is abstract and not suitable as a conclusive statement in research. Conclusions should be expressed with an indicator that is convincing, concretely understandable, and allows the effect to be felt sincerely. In this chapter, we introduce a generative quantity that makes this possible.

6.1 Inference of the Difference Between Two Independent Groups

Table 6.1 shows the raw data for the "Hospital Stay Problem" introduced in Sect. 3.1. The hospital stay of the ith patient in the experimental group is denoted as $x_{\text{exp}i}$, and that of the ith patient in the control group as $x_{\text{con}i}$.

6.1.1 Modeling the Data Generation Process

Assuming that the hospital stay of the experimental group follows a normal distribution with population mean μ_{exp} and population standard deviation σ_{exp}, the prob-

H. Toyoda, *Statistical Significance and the PHC Curve*,
https://doi.org/10.1007/978-981-97-7748-8_6

ability of observing a hospital stay of $x_{\mathrm{exp}1} = 5.5$ for the first patient[1] is expressed as:

$$L_{\mathrm{exp}1} = \mathrm{Normal}(x_{\mathrm{exp}1} = 5.5 | \mu_{\mathrm{exp}}, \sigma_{\mathrm{exp}}). \tag{6.1}$$

Similarly, the probability of observing the hospital stay $x_{\mathrm{exp}i}$ for the ith patient is

$$L_{\mathrm{exp}i} = \mathrm{Normal}(x_{\mathrm{exp}i} | \mu_{\mathrm{exp}}, \sigma_{\mathrm{exp}}), (i = 1, \ldots, n_1) \tag{6.2}$$

where $n_1 = 24$.

Assuming that the hospital stays are independent among patients, the simultaneous probability of observing the hospital stay data for the 24 patients in the experimental group

$$x_{\mathrm{exp}} = (x_{\mathrm{exp}1}, \ldots, x_{\mathrm{exp}i}, \ldots, x_{\mathrm{exp}24}) = (5.5, 10.0, 5.5, \ldots, 4.0, 9.0)$$

is expressed, given the normal distribution parameters $\mu_{\mathrm{exp}}, \sigma_{\mathrm{exp}}$, as

$$f(x_{\mathrm{exp}} | \mu_{\mathrm{exp}}, \sigma_{\mathrm{exp}}) = L_{\mathrm{exp}1} \times \cdots \times L_{\mathrm{exp}i} \times \cdots \times L_{\mathrm{exp}24}. \tag{6.3}$$

Similarly, assuming that the hospital stay of the control group follows a normal distribution with parameters $\mu_{\mathrm{con}}, \sigma_{\mathrm{con}}$, the simultaneous probability of observing the hospital stay data x_{con} for the 26 patients in the control group under the same assumption is

$$f(x_{\mathrm{con}} | \mu_{\mathrm{con}}, \sigma_{\mathrm{con}}) = L_{\mathrm{con}1} \times \cdots \times L_{\mathrm{con}i} \times \cdots \times L_{\mathrm{con}n_2} \tag{6.4}$$

where $n_2 = 26$.

6.1.2 Bayes' Theorem

Here we restate Bayes' Theorem (Eq. 5.22)

$$f(\theta | x) \propto f(x | \theta) f(\theta).$$

For the inference of the difference between two independent groups, we replace θ and x in this equation with

$$\theta = (\mu_{\mathrm{exp}}, \sigma_{\mathrm{exp}}, \mu_{\mathrm{con}}, \sigma_{\mathrm{con}}) \tag{6.5}$$

$$x = (x_{\mathrm{exp}}, x_{\mathrm{con}}) \tag{6.6}$$

[1] Strictly speaking, this is a probability density, but here we refer to it simply as probability.

Table 6.1 Raw data of hospital stay duration for patients administered antibiotics A and B (unit: days)

Experimental group (Group A) $n_1 = 24$	05.5, 10.0, 05.5, 06.0, 09.0, 09.5, 06.5, 07.0, 12.5, 07.0, 06.5, 10.5, 09.0, 04.5, 06.5, 09.5, 10.0, 09.5, 10.5, 06.5, 08.5, 12.5, 04.0, 09.0
Control group (Group B) $n_2 = 26$	05.5, 11.5, 07.0, 07.5, 10.5, 11.0, 08.0, 08.5, 14.0, 08.5, 10.0, 08.0, 12.0, 10.5, 06.0, 08.0, 11.0, 11.5, 11.0, 12.0, 08.0, 10.0, 14.0, 05.5, 10.5, 09.0

and use Bayes' Theorem. By specifying the likelihood and the prior distribution, the posterior distribution of the four parameters can be obtained using the MCMC method.

6.1.3 Likelihood

Since the measurements of the experimental and control groups are independent of each other, the simultaneous probability of observing x_{exp} and x_{con} is

$$f(x|\theta) = f(x_{\text{exp}}, x_{\text{con}}|\theta) = f(x_{\text{exp}}|\mu_{\text{exp}}, \sigma_{\text{exp}})f(x_{\text{con}}|\mu_{\text{con}}, \sigma_{\text{con}}). \tag{6.7}$$

The data is fixed to the values in Table 6.1. Therefore, treating this equation's data as constants and the parameters as variables, it becomes the likelihood.

6.2 Breaking Free from Zombies

By abandoning significance testing, we can be freed from the zombie problem. We discard the concept of a sampling distribution and use the likelihood principle, which is independent of intent. The simultaneous probability of observing x_{exp} and x_{con} in Eq. (6.7) does not depend on intents 1, 3, 4.

Whether we determine the number of observations in advance, set a period in advance, or decide to monitor the data several times along the way and calculate the PHC each time, the probability of observing the given data is described by Eq. (6.7). It is uniquely determined, regardless of intent. You are free to increase the number of subjects or calculate interim results as many times as you like, making prior registration unnecessary.

If prior registration can be omitted, you are free to suddenly change variables, stimuli, conditions, subjects, or n. The agony of creation is a series of trial and error. Changes, changes are needed. For creative discovery, methods based on the likelihood principle are more suitable than significance testing.

6.3 Prior Distribution

When substituting Eq. (6.5) for θ in Bayes' theorem, the prior distribution is the joint distribution of the four parameters. Normally, it is assumed that the parameters are independent of each other before observing the data, and it is expressed as the product of the prior distributions of the four parameters:

$$f(\theta) = f(\mu_{\text{exp}}, \sigma_{\text{exp}}, \mu_{\text{con}}, \sigma_{\text{con}}) \tag{6.8}$$

$$= f(\mu_{\text{exp}}) f(\sigma_{\text{exp}}) f(\mu_{\text{con}}) f(\sigma_{\text{con}}). \tag{6.9}$$

6.3.1 Uniform Distribution

In this book, we use a uniform distribution as the prior distribution. The prior distribution has already appeared in Chap. 5. Here, we introduce the precise formula for the distribution. A uniform distribution with an upper limit β and a lower limit α is expressed as

$$\text{Uniform}(y|\alpha, \beta) = \begin{cases} 1/(\beta - \alpha), & \alpha \leq y \leq \beta \\ 0, & \text{otherwise} \end{cases} \tag{6.10}$$

y takes continuous values, and if a value outside the range $\alpha \leq y \leq \beta$ is substituted, it returns 0.

In Chap. 5, we used a uniform distribution in the interval [0, 1] as the prior distribution for the population ratio π. In that case, $\beta = 1$ and $\alpha = 0$, so actually,

$$f(\pi) = \text{Uniform}(\pi|\alpha = 0, \beta = 1) = \frac{1}{1-0} = 1. \tag{6.11}$$

Since the population ratio is defined in the interval [0, 1], there is no need to consider values outside this range, and $f(\pi) = 1$ always holds. Substituting this into Eq. (5.25) results in Eq. (5.26). As a result, the influence of the prior distribution disappears from the posterior distribution.

6.3.2 *Determining the Parameters of the Uniform Distribution*

When using a uniform distribution as the prior distribution for the parameters, how should we determine its lower limit α and upper limit β? In the "Hospital Stay Problem," there are a total of four parameters: two population means $\mu_{\text{exp}}, \mu_{\text{con}}$ and two population standard deviations $\sigma_{\text{exp}}, \sigma_{\text{con}}$. This means we need to determine four lower limits and four upper limits. There are three main methods for this.

The first method is to determine based on the nature of the observed variable x. For example, in the "Hospital Stay Problem," the observed variable x is the duration of hospital stay. Unlike temperature or profit and loss, hospital stay duration never takes negative values. Therefore, the lower limits of the prior distributions for μ_{exp} and μ_{con} can be set to 0. The same applies if the unit of the observed variable is weight or length.

The second method is to determine based on the mathematical properties of the parameters. The standard deviation is the square root of the average of the squares of deviations from the mean. Mathematically, the average of the squares of deviations can never be negative. Therefore, the lower limits of the prior distributions for σ_{exp} and σ_{con} can be set to 0. This can be done regardless of the unit of the observed variable.

The third method is to determine them arbitrarily. There are ways to determine them arbitrarily without affecting the results, so rest assured. Since the lower limit α has already been determined, here we take the upper limit β as an example. Specify a value in a region that is implausibly high for an average or variance.

6.3.3 *The Upper and Lower Limits of the Uniform Distribution Can Be Determined Loosely*

Let's consider a concrete example. A person with medical knowledge might fairly speculate, "What could be the upper limits of the average and standard deviation of hospital stay duration for patients targeted by the antibiotic in question?" The maximum value in Table 6.1 is 14 days, and the average is between 8 and 10 days, but one should not be swayed by these figures. Suppose worldwide experts agree that no one would realistically claim "the average hospital stay duration is over 20 days." In this case, you might set the upper limit β a bit generously, say at 50 days or even 100 days. If you're concerned, setting β at 500 days is also fine.

The standard deviation in Table 6.1 is about 2.3 days, but again, one should not be swayed by this figure. Suppose worldwide experts agree that no one would realistically claim "the average variability in hospital stay duration is over 10 days." In this case, you might set the upper limit β a bit generously, say at 25 days or even 50 days. If you're concerned, setting β at 100 days is also fine.

In short, the upper limit β should be set by considering "what would be a generously high yet plausible lower limit for an impossibly large value." (Although it was not necessary in this case) If you ever need to arbitrarily determine a lower limit α, you would consider "what would be a generously low yet plausible upper limit for an impossibly small value."

Let's explain why this approach is acceptable below.

6.3.4 Unique Properties of the Uniform Distribution as a Prior

From the aforementioned considerations, let's assume we chose a uniform distribution with $\alpha = 0, \beta = 100$ as the prior distributions $f(\mu_{\mathrm{exp}}), f(\sigma_{\mathrm{exp}}), f(\mu_{\mathrm{con}}), f(\sigma_{\mathrm{con}})$.[2] Then, in the interval [0, 100], regardless of the value of y, we have

$$\mathrm{Uniform}(y|\alpha = 0, \beta = 100) = \frac{1}{100 - 0} = 0.01. \tag{6.12}$$

Therefore, the joint prior distribution (6.9) becomes a constant when all four variables are within the interval [0, 100]:

$$f(\theta) = 0.01 \times 0.01 \times 0.01 \times 0.01 = 0.01^4, \tag{6.13}$$

and Bayes' theorem simplifies to

$$f(\theta|x) \propto f(x|\theta) \times 0.01^4 \tag{6.14}$$

$$\propto f(x|\theta). \tag{6.15}$$

The posterior distribution is influenced only by the likelihood, and the effect of the prior distribution disappears. In other words, the posterior distribution is composed solely of the data information in the likelihood. As we learned in Chap. 5, the uniform distribution as a prior distribution was non-informative in this sense.

What if the values of the parameters fall outside the interval [0, 100]? For example, if the value of μ_{exp} is 200, at least one factor on the right side of (6.13) becomes 0. Since it's a product, the entire joint prior distribution $f(\theta)$ becomes 0. However, the likelihood of an implausible population mean like $\mu_{\mathrm{exp}} = 200$ is also nearly 0. Therefore, even outside the range, Eq. (6.15) holds true almost entirely, and the posterior distribution is almost proportional to the likelihood.

[2] The upper limits of the prior distributions for the population mean and standard deviation can differ from each other. Here, for simplicity and without affecting the results, we set them uniformly to 100. However, for example, the upper limit for the prior distribution of μ_{exp} could be 100, and for σ_{exp} it could be 50. But the upper limits for the prior distributions of the experimental and control groups must be the same. For instance, it would not be appropriate to have an upper limit of 20 for μ_{exp} and 100 for μ_{con}. The conditions for the two groups being compared should be equal.

The important thing is that the lower and upper limits of the prior distribution cover the main region of the likelihood function. As long as they cover it, whether the coverage is small or large, the posterior distribution is proportional to the likelihood. This property, that the posterior distribution is unaffected whether the coverage is small or large, is very convenient when determining the lower and upper limits. This is because specifying a generously high yet plausible lower limit for an impossibly large value, or a generously low yet plausible upper limit for an impossibly small value, essentially does not affect the posterior distribution.[3] The analysis results presented in this book are essentially invariant.[4]

In simpler terms, the analysis results are essentially invariant as long as the range is not too narrow. This flexibility in determining the range of the uniform distribution is why software like STAN automatically and appropriately decides it without specification. The property that the analyst does not necessarily have to specify it is very convenient in practice. For elementary statistical models, the uniform distribution is recommended as a prior distribution.

6.4 Inference of the Difference Between Two Independent Groups

The likelihood was specified as Eq. (6.7), and the four prior distributions were specified as Eq. (6.12). Using both, we apply Bayes' theorem and specify the posterior distribution as shown in Eqs. (6.14) and (6.15). Once the posterior distribution is specified, we generate T random numbers like $\theta^{(1)}, \theta^{(2)}, \ldots, \theta^{(t)}, \ldots, \theta^{(T)}$ using the MCMC method to approximate the posterior distribution of the parameter θ. Here, let's set $T = 100{,}000$, and for example, generate random numbers as follows:

[3] For the prior distribution of the population mean, distributions like the normal distribution, t-distribution, Cauchy distribution, and for the standard deviation, distributions like the inverse gamma distribution, half-normal distribution, half-t distribution, half-Cauchy distribution, and others are candidates. However, these distributions commonly have an unknown parameter called the scale parameter. Unfortunately, the value of the scale parameter affects the results of Bayesian analysis, including PHC, so it cannot be determined arbitrarily. Nor is it easy to fix the value of the scale parameter with a basis. Also, prior distributions other than the uniform distribution leave traces in the posterior distribution. For beginners, the uniform distribution is recommended because it allows for a rough treatment and leaves no trace in the posterior distribution. However, there are situations where various prior distributions are necessary due to various circumstances, and advanced users utilize them.

[4] However, the Bayesian factor, not covered in this book, is sensitive to the range of the uniform distribution.

$$\theta^{(1)} = \left(\mu_{\text{exp}}^{(1)}, \mu_{\text{con}}^{(1)}, \sigma_{\text{exp}}^{(1)}, \sigma_{\text{con}}^{(1)}\right) = (8.44, 9.42, 2.05, 2.34) \tag{6.16}$$

$$\theta^{(2)} = \left(\mu_{\text{exp}}^{(2)}, \mu_{\text{con}}^{(2)}, \sigma_{\text{exp}}^{(2)}, \sigma_{\text{con}}^{(2)}\right) = (7.97, 9.82, 2.19, 2.49) \tag{6.17}$$

$$\theta^{(3)} = \left(\mu_{\text{exp}}^{(3)}, \mu_{\text{con}}^{(3)}, \sigma_{\text{exp}}^{(3)}, \sigma_{\text{con}}^{(3)}\right) = (8.47, 9.96, 2.11, 2.50) \tag{6.18}$$

$$\vdots \qquad \vdots \qquad \vdots$$

$$\theta^{(99{,}999)} = \left(\mu_{\text{exp}}^{(99{,}999)}, \mu_{\text{con}}^{(99{,}999)}, \sigma_{\text{exp}}^{(99{,}999)}, \sigma_{\text{con}}^{(99{,}999)}\right) = (7.67, 9.69, 2.36, 2.35)$$

$$\theta^{(100{,}000)} = \left(\mu_{\text{exp}}^{(100{,}000)}, \mu_{\text{con}}^{(100{,}000)}, \sigma_{\text{exp}}^{(100{,}000)}, \sigma_{\text{con}}^{(100{,}000)}\right) = (8.65, 9.96, 2.42, 2.25).$$

Of course, since these are random numbers, the values will differ depending on the seed. This is just an example and does not always occur this way. T is not limited to 100,000.[5] In the case of complex models created by advanced users, it could be in the thousands.

6.4.1 EAP, post.sd, Credible Intervals

The point estimate of the parameter was obtained as the average of the random numbers. For example, for μ_{exp}, it is calculated as

$$8.15 = \frac{1}{100{,}000}(8.44 + 7.97 + 8.47 + 7.90 + 8.12 + \cdots + 7.67 + 8.65).$$

This is called the EAP estimate. The typical spread of the parameter was obtained as the standard deviation of the random numbers. For example, for μ_{exp}, it is calculated as

$$0.51 = \sqrt{\frac{1}{100{,}000}\left[(8.44 - 8.15)^2 + (7.97 - 8.15)^2 + \cdots + (8.65 - 8.15)^2\right]}.$$

This is called the posterior standard deviation or post.sd.

The interval estimate of the parameter was obtained as the credible interval. For example, the 2.5th percentile of the 100,000 $\mu_{\text{exp}}^{(t)}$ is 7.14, and the 97.5th percentile is 9.16. Thus, the 95% credible interval for μ_{exp} is [7.14, 9.16]. As we learned in Chap. 5, the summary statistics of the posterior distribution of the parameter μ_{exp} are expressed as 8.15 [7.14, 9.16]. We interpret this as the average hospital stay duration with the new antibiotic A being about 8.15 days, and with 95% credibility, the average hospital stay duration is between 7.14 and 9.16 days.

[5] For the models appearing in this book, $T = 100{,}000$ is fixed because it is sufficiently large for confidence. For those interested in how to determine T, please refer to the literature [8, 15].

The EAP estimate for σ_{exp} is the average value of the random numbers $\sigma_{\text{exp}}^{(t)}$, which was 2.48. The summary statistics of the posterior distribution of the parameter σ_{exp} are 2.48 [1.85, 3.39]. This means that the hospital stay with the new antibiotic A varies on average by about 2.48 days around the average hospital stay duration. The average variability in hospital stay duration is interpreted to be between 1.85 and 3.39 days with 95% credibility.

Please distinguish between the post.sd of μ_{exp}, which is 0.51 in the above formula, and the EAP estimate of σ_{exp}, which is 2.48. The former is the spread of the average hospital stay duration, and the latter is the spread of the hospital stay duration. To avoid complex interpretations, we will omit the interpretation of post.sd[6] going forward.

The summary statistics for the parameter μ_{con} were 9.58 (0.49) [8.61, 10.55], and for the parameter σ_{con}, they were 2.47 (0.37) [1.87, 3.32]. Please interpret these correctly.

6.5 Generative Quantities

Generally, a real function of the parameters, $g(\theta)$, is called a generative quantity. Therefore, PHC is also a generative quantity. The posterior distribution of the generative quantity $g(\theta)$ can be approximated by $g(\theta^{(t)})$. In simpler terms, the posterior distribution of a generative quantity can be approximated by applying its defining formula directly to the random numbers of the parameters generated by MCMC. This is a notable advantage of the Bayesian approach. In the following five sections, we will discuss considerations using generative quantities.

6.6 Posterior Distribution of the Difference in Population Means

To infer the difference in average values between the experimental and control groups, we consider the posterior distribution of the difference in population means, μ_{diff}. Since μ_{diff} is a function of the two population means, it can be approximated as

$$\mu_{\text{diff}}^{(t)} = \mu_{\text{con}}^{(t)} - \mu_{\text{exp}}^{(t)}. \tag{6.19}$$

For example, among the T values, as illustrated in Eqs. (6.16) and (6.17),

$$0.98 = \mu_{\text{diff}}^{(1)} = \mu_{\text{con}}^{(1)} - \mu_{\text{exp}}^{(1)} = 9.42 - 8.44 \tag{6.20}$$

$$1.85 = \mu_{\text{diff}}^{(2)} = \mu_{\text{con}}^{(2)} - \mu_{\text{exp}}^{(2)} = 9.82 - 7.97. \tag{6.21}$$

[6] post.sd provides an interpretation equivalent to the standard error in frequentist theory.

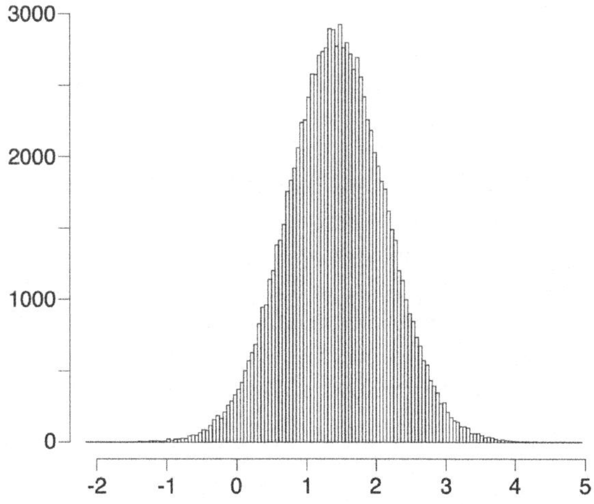

Fig. 6.1 Posterior distribution of the difference in population means

Please compare this with Eqs. (6.16) and (6.17). The histogram of $\mu_{\text{diff}}^{(t)}$ calculated for $T = 100{,}000$ is shown in Fig. 6.1, and this is the approximation of the posterior distribution of the difference in population means by random numbers.

The summary statistics of the posterior distribution of the difference in population means, μ_{diff}, are 1.43 [0.03, 2.82]. It is inferred that using antibiotic A reduces the average hospital stay by 1.43 days compared to using antibiotic B, and this average reduction in days is with 95% credibility between 0.03 and 2.82 days.

6.6.1 Probability that the Hypothesis of a Difference Is Correct

The probability that the research hypothesis, stating "using antibiotic A reduces the average hospital stay by more than c days compared to using antibiotic B," is correct can be calculated using Eq. (5.28) as

$$u_{c<\mu_{\text{diff}}}^{(t)} = \begin{cases} 1 & c < \mu_{\text{diff}}^{(t)}, \quad (t = 1, \ldots, T) \\ 0 & \text{otherwise} \end{cases} \tag{6.22}$$

by taking the average of $u_{c<\mu_{\text{diff}}}^{(t)}$. It is very important here not to use PHC ($0 < \mu_{\text{diff}}$) alone as the criterion for evaluating (peer review of) a paper. If we do, we fall back into the "invisible hand" trap, even after abandoning the sample distribution and freeing ourselves from intent. Even a slight improvement, such as discharging patients 5 min earlier, will make PHC ($0 < \mu_{\text{diff}}$) approach 1.0 as the number of observations increases. Judging the value of research based on this leads to the same misunderstandings as significance testing.

Table 6.2 Probability that the hypothesis of a difference in population means is correct (PHC ($c < \mu_{\text{diff}}$))

c	0.0	0.1	0.2	0.3	0.4	0.5	0.6	0.8	1.0
PHC	0.98	0.97	0.96	0.95	0.93	0.91	0.88	0.82	0.73
c	1.2	1.4	1.6	1.8	2.0	2.5	3.0	3.5	4.0
PHC	0.63	0.52	0.41	0.30	0.21	0.07	0.01	0.00	0.00

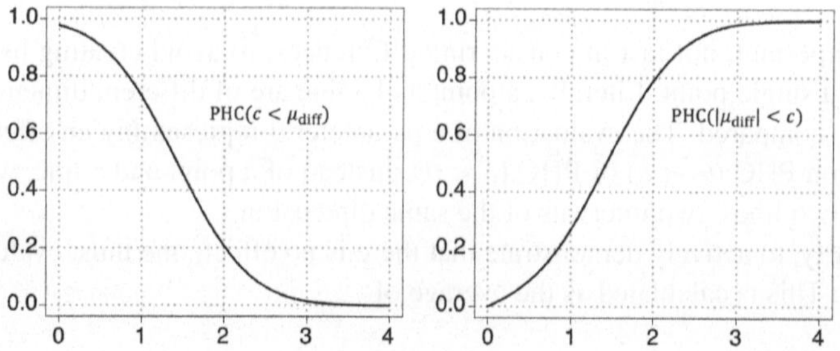

Fig. 6.2 PHC curve for the difference in population means

Furthermore, the evaluation of c is not a statistical problem. What can be done if patients are discharged on average one day (or one hour) earlier? Is a reduction of c days worth the additional costs of various antibiotic A treatments? This judgment is made by peer reviewers using domain knowledge.

Table 6.2 shows PHC ($c < \mu_{\text{diff}}$), varying c from 0.0 to 4.0 days in appropriate increments. Figure 6.2 on the left shows the PHC curve for this interval. There is over 90% credibility that switching to antibiotic A will result in an average discharge at least half a day earlier.

Analysts do not need to specify the value of c, either before or after the research. Researchers in drug development simply need to focus on shifting this curve to the right. They submit their work and leave the judgment to peer reviewers.

Reviewers do not make statistical judgments. They assess the academic value of this curve strictly from the perspective of patient convenience and cost.

The value judgment by reviewers will not be unanimous, but that is the essence of academic value judgment and peer review. Research activities are not so superficial that they can be judged based on whether a certain statistical indicator exceeds a certain value, regardless of the research context.

6.6.2 Probability that the Hypothesis of No Difference Is Correct

To argue that there is no difference between antibiotics A and B, one should not consider the research hypothesis that there is absolutely no difference. The probability that such a hypothesis is correct is 0 (equal to PHC ($\mu_{\text{diff}} = 0$)) even before collecting data. This would be repeating the mistake made by using the null hypothesis of no difference.

It is important, not just in considering differences, to avoid creating hypotheses based on a single point. Literally, a point and a line are of different dimensions and cannot be compared. The evaluation of a parameter θ representing an effect should be based on PHC ($\theta < c$) or PHC ($c < \theta$). Instead of a point and a line, we should compare two lines, two intervals of the same dimension.

Similarly, to actively demonstrate that there is no effect, one must evaluate PHC ($|\theta| < c$). This is calculated as the average of

$$u^{(t)}_{|\mu_{\text{diff}}|<c} = \begin{cases} 1 & |\mu^{(t)}_{\text{diff}}| < c, \quad (t = 1, \ldots, T) \\ 0 & \text{otherwise.} \end{cases} \tag{6.23}$$

Table 6.3 shows PHC ($|\mu_{\text{diff}}| < c$), varying c from 0.0 to 1.0 and from 2.0 to 3.0 in increments of 0.1. Figure 6.2 on the right shows the PHC curve for c ranging from 0.0 to 4.0. The fact that it is 0 (equal to PHC ($\mu_{\text{diff}} = 0$)) before collecting data is reflected in the left endpoint of this curve starting at 0. In this case, too, the analyst does not necessarily need to determine the value of c to a single point. It is sufficient to present Fig. 6.2 on the right and Table 6.3 and leave the evaluation to the reader.

In my opinion, being able to discharge patients more than half a day earlier would be convenient for various reasons. This is purely an amateur and personal view, but I do not consider a difference of about a quarter of a day to be significant. From this perspective, the probability that the hypothesis of no substantial difference is correct is below 5%.

Conversely, to have more than 90% credibility in the hypothesis of no difference, one must take the position that an average early discharge of less than 2.5 days is essentially meaningless. I believe that a lot can be done in 2.5 days. It is impossible for me to take such a stance.

Table 6.3 Probability that the hypothesis of no difference in population means is correct (PHC ($|\mu_{\text{diff}}| < c$))

c	0.0	0.1	0.2	0.3	0.4	0.5	0.6	0.7	0.8	0.9	1.0
PHC	0.00	0.01	0.03	0.05	0.07	0.09	0.12	0.15	0.18	0.22	0.27
c	2.0	2.1	2.2	2.3	2.4	2.5	2.6	2.7	2.8	2.9	3.0
PHC	0.79	0.83	0.86	0.89	0.92	0.93	0.95	0.96	0.97	0.98	0.99

6.7 Standardized Mean Difference

The significance test for the difference between two independent groups is designed based on Intent 1, but when only the final results are reported based on Intent 4, even with open data, it is impossible to distinguish between the two. However, the true risk in the latter case often becomes unacceptably high. This contributes to the lack of reproducibility in research, necessitating power analysis before data collection to fix n and pre-register this fact.

For power analysis, an indicator of the degree of deviation from the null hypothesis was needed. For example, in the case of testing the difference between two independent groups, it is the standardized mean difference (Eq. (3.8)), which is restated and transformed as

$$\delta = \frac{\mu_{\text{con}} - \mu_{\text{exp}}}{\sigma_{\text{within}}} = \frac{\mu_{\text{diff}}}{\sigma_{\text{within}}}. \tag{6.24}$$

For power analysis, for instance, setting $\delta = 0.3$, $1 - \beta = 0.8$, determines the number of observations as $n_1 = n_2 = 176$.

6.7.1 Power Analysis May Not Always Be Valid

However, the standard deviation in the denominator is unknown before data collection. After power analysis and before data collection (at the time of pre-registration), the research goal for reducing hospital stay duration in the study is expressed as

$$\mu_{\text{diff}} = \delta \times \sigma_{\text{within}} = 0.3 \times \sigma_{\text{within}}. \tag{6.25}$$

If the standard deviation turns out to be one day after data collection, in retrospect, the research goal would have been just 0.3 days. Thus, the true research goal using δ is only revealed after data collection.

Conversely, if the standard deviation was 10 days, the research goal would become 3 days. A goal of 0.3 and 3 days differ significantly in medical terms, affecting the impact and significance of the research. At least from the patient's perspective, the value of δ is non-essential; what matters is how much earlier they can be discharged. For power analysis, δ must be chosen as the indicator of deviation from the null hypothesis. However, δ is not always valid as a measure of research effect size.

If pre-registration requires fixing δ to a single point and conducting power analysis before writing a paper, readers would be left wondering what would happen if δ had other values. Power analysis is as irrational as shipping an air conditioner that has its temperature setting fixed and unchangeable.

6.7.2 Posterior Distribution of the Standardized Mean Difference

When told that the difference in average hospital stay duration is 1.5 days, its meaning is immediately clear. However, if told that the difference in average scores between groups in a psychological test is 1.5 points, it's not immediately clear whether this difference is large or small. In such research where it's difficult to give concrete meaning to the raw difference in measurements, δ becomes effective. It allows for re-evaluating the mean difference in terms of the average spread within groups. Here, let's consider the indicator

$$\delta_{\text{con}} = \frac{\mu_{\text{con}} - \mu_{\text{exp}}}{\sigma_{\text{con}}}. \tag{6.26}$$

Since antibiotic B is an older drug with more information available, σ_{con} can be imagined by many doctors as a standard for the variability of hospital stay duration among patients.[7] Substituting the sample mean and standard deviation gives 0.652 (= $9.6 - 8.1/2.3$). The difference in average hospital stay duration due to the difference in antibiotics is interpreted as 65.2% of the average variability in hospital stay duration within the control group. Multiplying by 10 allows for interpretation as a deviation value.[8] Considering the distribution of the control group as a reference, the deviation value of the average difference in hospital stay duration is 6.52.

The posterior distribution of the mean difference standardized by the standard deviation of the control group can be approximated as

$$\delta^{(t)}_{\text{con}} = \frac{\mu^{(t)}_{\text{con}} - \mu^{(t)}_{\text{exp}}}{\sigma^{(t)}_{\text{con}}}. \tag{6.27}$$

For example, among the T values, as illustrated in Eqs. (6.16) and (6.17),

$$0.42 = (\mu^{(1)}_{\text{con}} - \mu^{(1)}_{\text{exp}})/\sigma^{(1)}_{\text{con}} = (9.42 - 8.44)/2.34 \tag{6.28}$$

$$0.74 = (\mu^{(2)}_{\text{con}} - \mu^{(2)}_{\text{exp}})/\sigma^{(2)}_{\text{con}} = (9.82 - 7.97)/2.49. \tag{6.29}$$

Please compare this with Eqs. (6.16) and (6.17). The histogram of $\delta^{(t)}_{\text{con}}$ calculated for $T = 100{,}000$ is shown in Fig. 6.3, and this is the approximation of the posterior distribution of the difference in population means by random numbers.

[7] Using σ_{exp} or a common standard deviation for both groups is also possible, but is omitted here due to space constraints.

[8] A deviation value difference of 10 means a difference of one standard deviation.

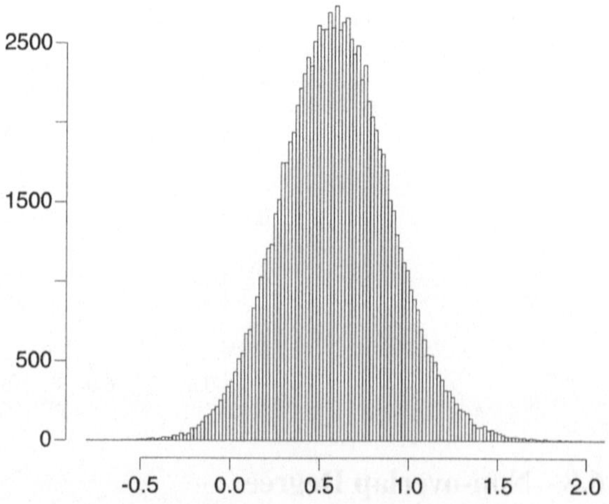

Fig. 6.3 Posterior distribution of the standardized mean difference

6.7.3 Probability that the Hypothesis $\delta_{con} > c$ Is Correct

The probability that the hypothesis stating "the standardized mean difference δ_{con} is greater than c" is correct can be calculated as

$$u^{(t)}_{c<\delta_{\text{con}}} = \begin{cases} 1 & c < \delta^{(t)}_{\text{con}}, \quad (t = 1, \ldots, T) \\ 0 & \text{otherwise} \end{cases} \tag{6.30}$$

by taking the average of $u^{(t)}_{c<\delta_{\text{con}}}$. Table 6.4 shows PHC ($c < \delta_{\text{con}}$), varying c from 0.0 to 0.4 in increments of 0.05. Additionally, Fig. 6.4 shows the PHC curve for the interval [− 0.5, 2.0]. Since a higher value is desirable for this indicator, the PHC curve slopes downward to the right. It seems likely that the average reduction in hospital stay is more than 20% of the average variability in hospital stay duration.

Analysts do not need to pre-determine not only the number of observations but also the benchmark c for δ_{con}. If reviewers request that "the standardized mean difference should be at least c", the probability that the research hypothesis "the standardized mean difference is greater than c" is correct can be shown in a revised manuscript. Alternatively, if the PHC curve is presented in the paper in advance, there is no need to even discuss a reasonable value of c with reviewers. In other words, there is no need to decide in advance, nor is there a need to decide after the analysis. However, for power analysis in significance testing, it was necessary to fix the standardized mean difference to a single point in advance, which is extremely inconvenient.

Table 6.4 Probability that the hypothesis $\delta_{\text{con}} > c$ is correct (PHC ($c < \delta_{\text{con}}$))

c	0.00	0.05	0.10	0.15	0.20	0.25	0.30	0.35	0.40
PHC	0.98	0.97	0.95	0.93	0.91	0.87	0.83	0.79	0.74

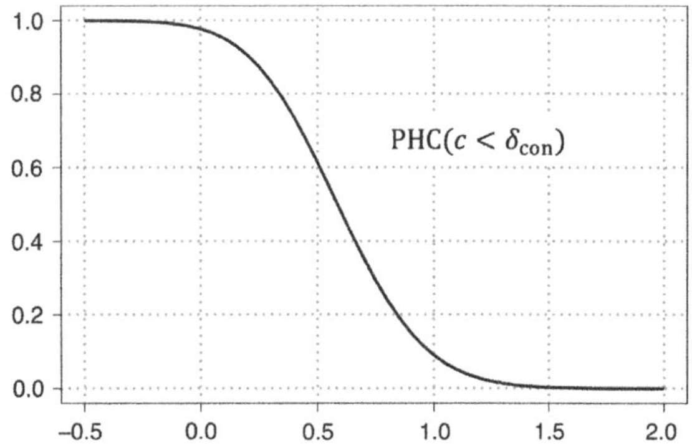

Fig. 6.4 PHC curve for the standardized mean difference

6.8 Non-overlap Degree

Recall the medical licensing exam question (Fig. 6.3) introduced in Chap. 2. Let's focus on the incorrect answer option d. Converting the percentage to probability c, it is restated as:

> c percent of patients in Group A have a shorter hospital stay than the average hospital stay in Group B.

In the question, it is stated as "96.4% of patients." It would be wonderful if many patients using the new drug could be discharged earlier than the average hospital stay with the old drug. Although it is an incorrect answer choice, it is more appropriate as a measure of the effect of the new drug than an uninterpretable p-value.

This indicator is the **non-overlap degree** proposed in the literature [10], defined as

$$U_3 = \{\text{cumulative probability up to } \mu_{con} \text{ of Normal } (\mu_{exp}, \sigma_{exp})\}. \tag{6.31}$$

This definition is complex, so Fig. 6.5 illustrates the non-overlap degree for clarity. The area of the hatched region is the non-overlap degree, representing the proportion of patients who can be discharged earlier than the average hospital stay in the old drug group. It is more heartfelt than a p-value.

By replacing $\mu_{exp}, \sigma_{exp}, \mu_{con}$ with random numbers, as before, we approximate the posterior distribution of U_3. The summary statistics of the posterior distribution of U_3 were 0.71 [0.50, 0.88]. Interpreted using the EAP estimate, "71% of patients in Group A have a shorter hospital stay than the average hospital stay in Group B." This is a heartfelt conclusion.

Following Eqs. (6.22), (6.23), and (6.30), we calculate the probability that the hypothesis "$U_3 > c$" is correct. Figure 6.6 shows the PHC curve for PHC ($c < U_3$) with c varying from 0.4 to 1.0. Since a higher value is desirable for this indicator, the PHC curve slopes downward to the right.

Table 6.5 shows the PHC values for c ranging from 0.4 to 1.0 in increments of 0.01. We conclude with 95% credibility that "at least 54% of patients in Group A

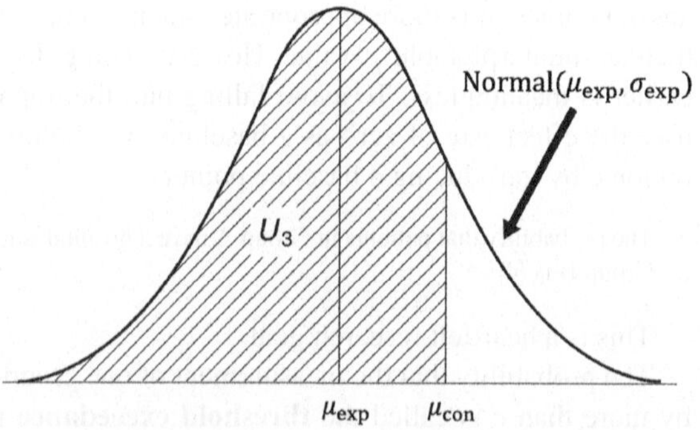

Fig. 6.5 Schematic diagram of non-overlap degree

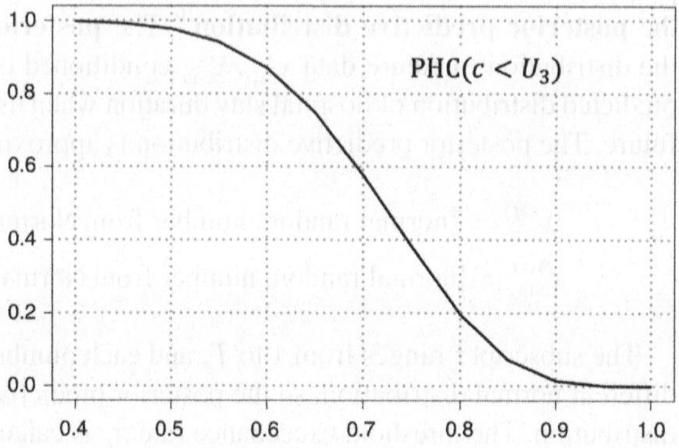

Fig. 6.6 PHC curve for non-overlap degree

Table 6.5 Probability that the hypothesis $U_3 > c$ is correct (PHC ($c < U_3$))

c	0.50	0.51	0.52	0.53	0.54	0.55	0.56	0.57	0.58	0.59	0.60
PHC	0.98	0.97	0.97	0.96	0.95	0.94	0.93	0.92	0.91	0.89	0.87

have a shorter hospital stay than the average hospital stay in Group B." However, since the change in the PHC curve is gradual, it would certainly be better to increase the number of observations.

6.9 Threshold Exceedance Rate

Now, let's focus on the incorrect answer option c in the medical licensing exam question (Fig. 6.3). Rewriting the specific percentage as b%:

> The probability that patients in Group A have a shorter hospital stay than those in Group B is b%.

It would be wonderful if patients using the new drug have a higher probability of being discharged earlier than those using the old drug. Although it is an incorrect answer choice, it is more appropriate as a measure of the effect size of the new drug than an uninterpretable p-value. However, being discharged just 5 min or even 3 s earlier is meaningless. To avoid falling into the trap of significance testing, which uses the effect size of zero as a baseline, we slightly modify the incorrect answer option c by introducing a baseline point c:

> The probability that patients in Group A have a hospital stay shorter by c days than those in Group B is $b\%$.

This is a heartfelt research goal.

The probability that the measurement of one group exceeds that of another group by more than c is called the **threshold exceedance rate π_c**. This probability was proposed in the literature [8]. Here, we introduce a simple evaluation method using the **posterior predictive distribution**.[9] The **posterior predictive distribution** is the distribution of future data $x^*_{\text{exp}}, x^*_{\text{con}}$ conditioned on the observed data. It is the predicted distribution of hospital stay duration when using antibiotics A and B in the future. The posterior predictive distribution is approximated as

$$x^{*(t)}_{\text{exp}} = \{\text{normal random number from Normal}\ (\mu^{(t)}_{\text{exp}}, \sigma^{(t)}_{\text{exp}})\} \tag{6.32}$$

$$x^{*(t)}_{\text{con}} = \{\text{normal random number from Normal}\ (\mu^{(t)}_{\text{con}}, \sigma^{(t)}_{\text{con}})\}. \tag{6.33}$$

The subscript t ranges from 1 to T, and each number is a random number from a different normal distribution, so the posterior predictive distribution is not a normal distribution. The threshold exceedance rate π_c is calculated as the average of

$$u^{(t)}_{c<x^*_{\text{con}}-x^*_{\text{exp}}} = \begin{cases} 1 & c < x^{*(t)}_{\text{con}} - x^{*(t)}_{\text{exp}}, \quad (t = 1, \ldots, T) \\ 0 & \text{otherwise.} \end{cases} \tag{6.34}$$

Figure 6.7 shows the threshold exceedance rate curve for the interval [0.0, 2.5]. Since a higher value is desirable for this indicator, the curve slopes downward to the right. Table 6.6 shows the threshold exceedance rate in the same interval at increments of 0.25. For example, the probability that the new drug leads to at least half a day ($c = 0.5$) shorter hospital stay than the old drug is 60% ($\pi_{0.5} = 0.60$).

Figure 6.8 explains how this is calculated. It is a scatter plot of the posterior predictive distribution. 100,000 random numbers generated by Eqs. (6.32) and (6.33) are plotted. The horizontal axis is x^*_{con}, and the vertical axis is x^*_{exp}. A line $x^*_{\text{exp}} = x^*_{\text{con}} - 0.5$ is drawn.

Transforming the condition in Eq. (6.34) gives $x^*_{\text{exp}} < x^*_{\text{con}} - 0.5$. Therefore, points to the right and below the line correspond to Eq. (6.34) being 1, and points to the left and above correspond to 0. In Table 6.6, $\pi_{0.5} = 0.6$ means that approximately 60% (about 60,000 points) of the points are in the right and below the area of the line.

---

[9] The literature [8] also shows how to obtain the posterior distribution of the threshold exceedance rate π_c. Using the posterior distribution, interval estimation of π_c for each c is also possible.

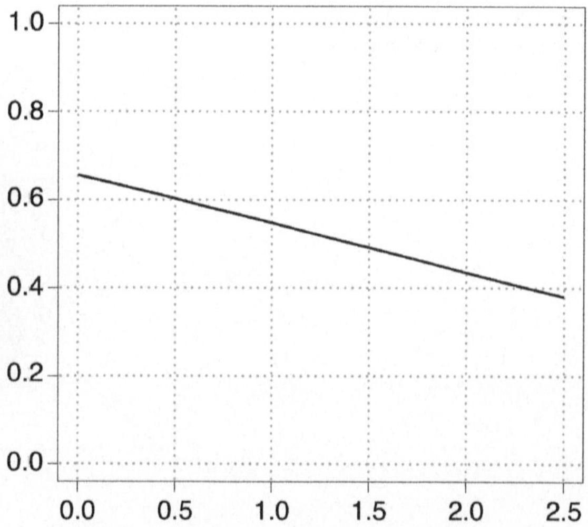

Fig. 6.7 Threshold exceedance rate curve

Table 6.6 Threshold exceedance rate for different c values

c	0.00	0.25	0.50	0.75	1.00	1.25	1.50	1.75	2.00	2.25	2.50
π_c	0.66	0.63	0.60	0.58	0.55	0.52	0.49	0.46	0.43	0.41	0.38

As the baseline point c increases, the line moves to the right. The number of points included in the right and below decreases, and the threshold exceedance rate π_c becomes smaller. For example, $\pi_{2.0} = 0.43$. The probability that the new drug leads to at least two days shorter hospital stay than the old drug is 43%.

Conversely, as the baseline point c decreases, the line moves to the left. The number of points included in the right and below increases, and the threshold exceedance rate π_c becomes larger. For example, $\pi_{0.25} = 0.63$. The probability that the new drug leads to at least a quarter of a day shorter hospital stay than the old drug is 63%. This is shown in Fig. 6.7 and Table 6.6.

6.10 Inference of Ratio

Let's focus on the incorrect answer option a in the medical licensing exam question (Fig. 6.3). Rewriting the specific percentage as $c\%$:

Group A has an average hospital stay that is $c\%$ shorter than Group B.

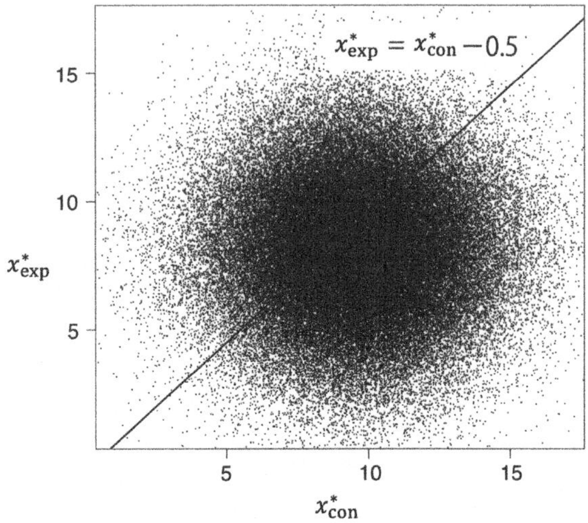

Fig. 6.8 Schematic diagram of threshold exceedance rate

Instead of confusing the point difference[10] with the percentage itself, let's rewrite it using the ratio notation, considering the average hospital stay of the control group as 1:

Group A has an average hospital stay that is c times shorter than Group B.

Unlike a p-value, this goal is understandable even to elementary school students and is heartfelt.

In cases where the measurement, like hospital stay duration, can be directly interpreted, the standardized mean difference is not always appropriate as a measure of research effect size. The mean difference in hospital stay duration itself is more appropriate. However, hospital stay duration, like weight or length, is an absolute quantity starting from 0. Being discharged 3 days earlier for a disease with an average one-year hospital stay and for a disease with an average 10-day hospital stay are entirely different. In the former case, the ratio is 0.99 (= 362/365), and in the latter, it is 0.70 (= 7/10). In some cases, the ratio is more scientifically important as a measure of research effect size than the mean difference.

The ratio of average hospital stay duration[11] For an appropriate calculation example, refer to Chap. 2 of the literature [8] being less than c is calculated as the average of

$$u^{(t)}_{\mu_{\mathrm{exp}}/\mu_{\mathrm{con}}<c} = \begin{cases} 1 & \mu^{(t)}_{\mathrm{exp}}/\mu^{(t)}_{\mathrm{con}} < c, \quad (t = 1, \ldots, T) \\ 0 & \text{otherwise.} \end{cases} \tag{6.35}$$

[10] For example, if the cabinet approval rating was 60% last time and 55% this time, the difference is said to be 5 points. A **point** is a unit of percent difference.

[11] The ratio of hospital stay duration can be obtained as the ratio of the posterior predictive distribution. In this example, we avoided calculating the posterior distribution of the ratio of hospital stay duration because a small number of negative values appeared in the posterior predictive values.

Table 6.7 Probability that the hypothesis of ratio being less than c is correct

c	0.90	0.91	0.92	0.93	0.94	0.95	0.96	0.97	0.98	0.99	1.00
PHC	0.76	0.80	0.84	0.87	0.90	0.92	0.94	0.95	0.96	0.97	0.98

The summary statistics of the ratio's posterior distribution were 0.85 [0.72, 1.00]. Table 6.7 shows PHC ($\mu_{\text{exp}}/\mu_{\text{con}} < c$), varying c from 0.9 to 1.0 in increments of 0.01. As a smaller value is desirable for this indicator, the probability increases with the increase in c. When considering the average hospital stay of the old drug as 1, the point estimate for the new drug is shortened to 0.85, and "the hypothesis that the average hospital stay for the new drug is less than 0.94" is correct with a 90% probability. The PHC curve for PHC ($\mu_{\text{exp}}/\mu_{\text{con}} < c$) is left as an exercise for the reader to draw, and is intentionally not shown here. Please try drawing it yourself.

When inferring the difference between two groups, non-overlap degree and threshold exceedance rate were easier to interpret than a p-value. Of course, not limited to "hospital stay duration" discussed in this chapter, non-overlap degree and threshold exceedance rate are easier to interpret than a p-value for any variable. For example, consider a diet experiment. If the non-overlap degree is 0.3, it means that the average weight of the group that followed the diet method is at the 30% point of the distribution of the group that did not. The effect of the diet within the group becomes visually clear. If the threshold exceedance rate with a baseline of 2 kg is 0.7, it means that 70% of the people who followed the diet method are at least 2 kg lighter than those who did not. The meaning is clear. Please try to concretely interpret non-overlap degree and threshold exceedance rate for typical variables in your field of expertise.

Interlude: Resolved Issues from Part I

The multiple incidents that occurred in Part I were resolved as follows.

1. **"Statistically Significant" Is Merely a Necessary Condition**
The previous peer review policy, which automatically deemed a study worthy of publication, irrespective of the research field, domain, context, or analysis purpose, and regardless of the subject matter or weight of the results, as long as it met the criterion of a p value less than 0.05, is incorrect. This approach should be discontinued. Instead, research questions should be flexibly set according to the specific circumstances of the research purpose, and responses should be given sincerely. For example:
The average hospital stay for Group A is more than half a day shorter than that for Group B. (Measured value)
71% of patients in Group A have a shorter hospital stay than the average for Group B. (Percentage, measure of non-overlap)

The probability that Group A will be discharged more than half a day earlier than Group B is 60%. (Probability, probability beyond threshold)
Compared to Group B, Group A has an average hospital stay ratio of 0.94 or less. (Ratio)
Such are the examples. The expression of conclusions should be changed to one that can be intuitively understood by domain experts, demonstrating a high probability that the statement is correct. It's not just about meeting a necessary condition but about contributing to the development of the application field by stating, with high probability, conditions that are sufficient. This is akin to only allowing athletes who have broken the standard records set for each event to participate in the Olympics.

2. The Invisible Hand of Research

When a researcher refrains from increasing n, the probability that their desired hypothesis (PHC) is correct does not increase significantly. As a result, the instability caused by a small n reduces the risk of an analyst's erroneous claims being accepted. Since the burden of proving the value of research primarily lies with the analyst, this feature is quite rational. As n increases, the curve becomes steeper, which means the PHC at any given c approaches 0 or 1. Therefore, with a larger n, clear judgments can be made irrespective of the analyst's claims or position. Data is information. The more appropriate data there is, the better. This is where the difference lies from the peculiar nature of significance tests that can be "statistically significant but academically meaningless with too much data." However, considering PHC with the reference point at zero (or no effect) can lead to the invisible hand of God.

3.1. Unable to Actively Show No Difference

To claim that there is no substantial difference, one should not consider a research hypothesis that there is absolutely no difference, as the probability that this hypothesis is correct, PHC $(\mu_{\text{diff}} = 0) = 0$, is zero even before collecting data. This would be repeating the mistake made by using the null hypothesis to claim no difference. Whether it's losing 1 pg of weight or being discharged 3 s earlier, these are not substantial differences. The important thing is not to create hypotheses about points. Literally, since points and lines are different dimensions, they cannot be compared. The probability that the hypothesis of no substantial difference is correct can be shown by PHC $(|\mu_{\text{diff}}| < c)$. Instead of points and lines, let's compare probabilities assigned to two intervals, two lines of the same dimension. Analysts do not need to specify the value of c as a specific point. By showing the PHC curve, they can indicate the extent to which there is no substantial difference.

3.2. No Need to Determine n Before Data Collection

Using PHC, analysts do not need to determine n in advance. Nor do they need to specify the value of the reference point c before or after collecting data. This is because it suffices to show the PHC curve that aggregates and visualizes the relationship between c and PHC. By observing the PHC curve, a comprehensive assessment of the value of the metric on the horizontal axis is possible.

3.3. The "Effect Size" in Power Analysis Is Not Always Valid as a Measure of Effect

Using generative quantities, it is possible to calculate probabilities like "the probability of losing 2 kg," as well as posterior distributions such as measure of non-overlap, probability beyond threshold, and ratio. We can prioritize real-world demands and freely choose indicators to evaluate the effects of research according to its purpose. Their meanings are clear, and above all, it is convenient that they do not need to be predetermined.

Medicine and psychology are supposed to be disciplines that explain and predict individual behavior. However, research paradigms that focus solely on significant differences in population means neglect the explanation and prediction of individual behavior itself. Probability beyond threshold, on the other hand, can show treatment differences for individual patients or humans, rather than group differences.

Column: The Multiplicity Problem

Making judgments about multiple parameters or generative quantities, or recalculating PHC multiple times while changing the reference point, is nothing but re-observing a single simultaneous posterior distribution in various ways. Therefore, no multiplicity problem arises. It's perfectly acceptable to reanalyze to test hypotheses conceived after seeing the data or even after the analysis. This is because the simultaneous posterior distribution already exists independently of inspiration or intention and does not change due to them. Reanalysis is merely a way of looking at the simultaneous posterior distribution from a different perspective, so it's not a violation of rules. Rather, it allows for the patient search for hidden treasures named insights that are already embedded in the simultaneous posterior distribution.

4. The Zombie Problem

By abandoning the concept of a sample distribution dependent on intention, we are freed from the zombie problem. Instead, we use the likelihood principle, which does not depend on intention. Whether the number of observations is predetermined, or the period is set in advance, or it's decided to monitor the data several times along the way and calculate PHC each time, the probability of the observed data occurring remains constant. The likelihood is uniquely determined, independent of intention. It's fine to increase the number of observations along the way, or to recalculate intermediate results as many times as you like, which means that pre-registration itself becomes unnecessary. If pre-registration can be omitted, variables, stimuli, conditions, subjects, or n can be changed suddenly. It's free to interrupt if it doesn't seem to be working well or, conversely, to increase n if it seems to be going well. The pains of creation are a series of trial and error. Changes, changes are needed. For creative discoveries, Bayesian methods that allow for free change are more suitable than significance tests.

Part III
Education and Malice

Chapter 7
Seligman's Dogs

Using the Difference of Corresponding Two Groups as an Example.

7.1 The Diet Problem Revisited

> **Diet Problem 2**: To investigate the effectiveness of a certain diet method, a company developing a one-month weight loss program recruited 30 participants. Is this diet method effective?

The participants' weight before and after the program, their height, and the difference in weight before and after are shown in Table 7.1. Summary statistics are shown in Table 7.2.

7.1.1 Test for the Difference Between Two Corresponding Groups

First, we will analyze this data using the test for the difference between two corresponding groups, which we learned in Chap. 2.

Step 1: Set the null hypothesis H_0 and the alternative hypothesis H_1.

Null hypothesis H_0: $\mu_{x_b - x_a} = 0$, Alternative hypothesis H_1: $\mu_{x_b - x_a} \neq 0$.

Here, $\mu_{x_b - x_a}$ represents the population mean of weight loss.

H. Toyoda, *Statistical Significance and the PHC Curve*,
https://doi.org/10.1007/978-981-97-7748-8_7

Table 7.1 Difference in weight before and after participation in the diet (kg)

Subject number	1	2	3	4	5	6	7	8	9	10
Weight before	48.5	45.6	56.9	52.6	52.6	60.5	52.2	55.3	58.8	54.0
Weight after	47.0	42.6	54.4	51.0	51.0	58.3	50.4	54.4	57.6	51.5
Height	148.5	147.4	164.6	158.3	158.3	169.7	154.0	162.3	170.5	164.3
Weight difference	1.5	3.0	2.5	1.6	1.6	2.2	1.8	0.9	1.2	2.5
Subject number	11	12	13	14	15	16	17	18	19	20
Weight before	47.3	57.4	48.2	49.9	47.9	54.1	55.1	49.7	48.7	53.6
Weight after	45.8	55.7	46.9	48.1	45.0	51.9	53.2	47.0	46.0	51.6
Height	150.1	165.3	148.0	154.1	151.0	156.8	162.0	153.8	152.3	159.8
Weight difference	1.5	1.7	1.3	1.8	2.9	2.2	1.9	2.7	2.7	2.0
Subject number	21	22	23	24	25	26	27	28	29	30
Weight before	52.0	42.0	58.0	49.6	54.8	50.3	49.3	51.9	46.0	53.2
Weight after	50.5	40.0	56.4	47.2	53.0	48.0	47.0	49.4	44.8	50.9
Height	157.4	146.9	162.4	153.7	161.5	154.8	149.7	161.1	148.0	155.5
Weight difference	1.5	2.0	1.6	2.4	1.8	2.3	2.3	2.5	1.2	2.3

Table 7.2 Summary statistics of data for 30 participants in the weight loss program

	Mean	Standard deviation	Median
Before weight (x_b) (kg)	51.9	4.27	52.1
After weight (x_a) (kg)	49.9	4.40	50.5
Height (x_h) (cm)	156.7	6.66	156.2
Weight loss ($x_b - x_a$) (kg)	2.0	0.54	1.9

Step 2: Assuming H_0 is true, calculate the test statistic.

Assuming that $x_b - x_a$ follows a normal distribution, the test statistic calculated using Eq. (2.9)

$$z = \frac{\bar{x}_{\text{difference}}}{s_{\text{difference}}} \times \sqrt{n} = 3.7 \times \sqrt{30} = 20.1 \tag{7.1}$$

can be approximated by the standard normal distribution.[1] Equation (3.9) is an indicator of deviation from the null hypothesis and was used in power analysis. The value 3.7 (= 2.0/0.54) in the above equation is the estimated value based on the data in Eq. (3.9). The average weight loss was estimated to be 3.7 times the average spread of weight loss.

Step 3: Calculate the p-value from the sample distribution.

p-value = 2.2×10^{-16}, 95% credible interval is [1.8 kg, 2.2 kg] was obtained.

Step 4: Reject or accept the null hypothesis H_0.

[1] Precisely, it's a t-distribution with degrees of freedom $df = 30 - 1$.

With p-value < 0.05, we reject the null hypothesis H_0 and determine that there is a statistically significant difference in weight loss.

The extent of deviation from the null hypothesis, which is essential for power analysis, was estimated post-analysis to be 3.7 times, but it was not predictable before the analysis. Also, the number of participants in the monitor was unforeseeable, so designing n for power analysis was not possible. Consequently, this study cannot be submitted to journals that require pre-registration of power analysis.

Regardless of the possibility of publication, what about the issue of multiplicity? Fortunately, the lab where the data, prohibited from being taken out, was stored had surveillance cameras. Therefore, it was possible to prove post hoc that no exploratory testing was conducted.[2] There is no multiplicity issue related to timing. Thus, this test is not a "Type 4 Zombie" by p-value monitoring. Moreover, since the test is conducted only once in the paper, there is no multiplicity issue regarding the level, attributes, or reference variables.

However, unfortunately, $n = 30$ was not predetermined but rather the number of people who happened to apply during the period. It's possible that there could have been 20 or 100 applicants. Therefore, this test is a "Type 3 Zombie" due to resource fixation. As a result, Eq. (7.1) follows an unknown mixed distribution, and the original sample distribution remains unclear. Hence, the credibility of the confidence interval is also questionable, and it should not be referred to in principle. But, keeping it a secret that this is a "Type 3 Zombie," Analyst A decided to report to the head of the development department.

7.1.2 Research Questions

Mr. A Director, a significant difference was observed. The p-value is 2.2×10^{-16}, which is extremely highly significant.

Director That's on the order of one in a quadrillion. Speaking of a quadrillion, it's more than Japan's national budget, but does such an extraordinary p-value mean a 30 kg weight loss in a month? Or does it guarantee a 100% success rate in weight loss?

Mr. A It's unrelated to the kg of weight loss or the probability of successful weight loss. The p-value cannot be interpreted.

Director We don't live on fantasies. Use interpretable metrics.

Mr. A Yes, the average weight loss is 3.7 times the average spread of weight loss.

Director Indeed, that reflects the nature of the diet method. However, the inferential statistical insights needed for market demand and effective advertising are not just that in this case.

[2] However, the footage must be preserved permanently, which is quite troublesome.

The Development Director was interested in inferential statistical insights for the following three **research questions** (RQ.). Research questions are also referred to as **research questions**.

RQ.1 Instead of Evaluating the Effect Size of Weight Loss with Eq. (3.9), **Please Use Eq.** (3.8)
The desire to lose weight and become beautiful is primarily based on relative comparison with others. However, the denominator in Eq. (3.9) focuses on the individual's variability in the difference in weight before and after. Information that the average weight loss is 3.7 times this is irrelevant to that desire. The effect of weight loss must be considered in comparison with people who did not undergo the weight loss method, as it's meaningless as a product otherwise. The real value of losing weight is being slimmer than your peers of the same age who did not use this method.

Therefore, the effect of weight loss should be evaluated not with Eq. (3.9) but with Eq. (3.8). Here, we should adopt (average weight loss/within-group standard d$eviation$) as the effect of the diet method and provide its inferential statistical insights.

RQ.2 Evaluate Weight Loss in Measurable Units Positively
It is known that many people are advised by doctors during health check-ups to lose weight for the prevention of lifestyle-related diseases. Moreover, they are often advised to aim for a weight loss of 1 kg per month. If we target this group of potential lifestyle-related disease patients, it could be profitable. However, both the relative Eqs. (3.8) and (3.9), which use the ratio to variability, are not directly relevant as measures of weight loss effectiveness in this case. We want to use the measurement unit itself in advertising, saying, "By participating in this diet program, you can lose more than 1 kg in a month." However, if the advertising copy is false, we could be sued by consumers. Please demonstrate that the claim of losing more than 1 kg is statistically valid.

RQ.3 Evaluate Weight Loss in Terms of Obesity Level
The effort and health implications of a 1 kg weight loss in a tall participant compared to a short participant, as well as the impact on appearance, are entirely different. Please don't mix them up. The main goal of the weight loss method is to reduce obesity levels. Therefore, evaluate its effect statistically not just in kg but also using the obesity level BMI.

7.1.3 In the Case of a Learner Who Memorized Test Statistics Without Understanding

Mr. A studied psychological statistics for 4 credits in college and wrote his thesis and dissertation using significance tests. However, he did not learn about modeling with likelihood or Bayesian inference. He memorized the test statistics without deriving them and did not learn their mathematical derivation. In this case, Mr. A's monologue would probably be as follows:

RQ.1 Instead of Evaluating the Effect Size of Weight Loss with Eq. (3.9), **Please Use Eq.** (3.8)

For me, statistics was a subject of memorizing the association between data formats and testing methods. Data from a within-subject design[3] should be analyzed using the test for the difference between two corresponding groups. I was strictly warned not to use the test statistic for the difference between two independent groups on data with correspondence. Data from a within-subject design can only be considered using Eq. (3.9).

RQ.2 Evaluate Weight Loss in Measurable Units Positively

Testing is aimed at rejecting the null hypothesis. I only learned to negate a negative state where there is no effect at all. Even if it's a common phrase used in health check-ups, I don't think I can calculate the probability that the positive research hypothesis "this weight loss method can make you lose more than 1 kilo" is correct. It is impossible for me to statistically evaluate the positive effect of research using measurement units.

RQ.3 Evaluate Weight Loss in Terms of Obesity Level

Since I only memorized the sample distribution of the test statistic without understanding, I never considered what the distribution of obesity levels would be. If the weight difference follows a normal distribution, BMI, which is a function of division by the square of height, probably does not follow a normal distribution. Maybe if I consult people who are knowledgeable about statistics or look up in books, I can find a test method for obesity levels? If I can't find a test method for obesity levels, there's nothing else to do but give up. For me, statistics was about memorizing procedures, and there was no room for improvisation.

7.2 Likelihood Modeling

Mr. B studied psychological statistics for 4 credits in college, where he learned modeling with likelihood and Bayesian inference to write his thesis and dissertation. However, he did not learn the principles behind the derivation of the MCMC method, merely memorizing that it is a method to approximate the posterior distribution. Despite not knowing the principles of the MCMC method, Mr. B can conduct the following analysis.

The foundation of data analysis, as learned, is to consider the data generation process. This is the starting point for all analyses. In Table 7.1, there appear to be four variables. However, since weight loss $= x_{\mathrm{b}} - x_{\mathrm{a}}$, there are effectively only three variables. When dealing with more than three variables in quantitative data, it is usual to assume a multivariate normal distribution as the first choice. Therefore, we assume that the ith subject $x_i = (x_{\mathrm{b}i}, x_{\mathrm{a}i}, x_{\mathrm{h}i})$ is generated from a trivariate normal distribution

[3] Data from two corresponding groups, when the observation subject is people, is also called within-subject design data.

$$f(x_i|\theta) = \text{multi\_normal}_3(x_i|\mu, \Sigma) \tag{7.2}$$

The data is already obtained, and the parameters are unknown, so we regard this equation as the likelihood. Here, μ is the population mean vector, and Σ is the population variance and covariance, with

$$\theta = (\mu, \Sigma) = (\mu_{\mathrm{b}}, \mu_{\mathrm{a}}, \mu_{\mathrm{h}}, \sigma_{\mathrm{b}}^2, \sigma_{\mathrm{a}}^2, \sigma_{\mathrm{h}}^2, \sigma_{\mathrm{ba}}, \sigma_{\mathrm{ah}}, \sigma_{\mathrm{bh}}) \tag{7.3}$$

Here, σ with two subscripts represents the population covariance between the variables indicated by the subscripts. Taking the total product over subscript i as in Eq. (6.3) gives us the likelihood $f(x|\theta)$.

7.2.1 Freedom from the Multiplicity Problem

Whether x results from stopping data collection midway, increasing the data, or conducting exploratory analysis during data collection, the likelihood is invariant to intention. Therefore, pre-registration of the number of observations n is unnecessary. Even without surveillance cameras, we are freed from the multiplicity problem related to timing.

According to Bayes' theorem, the posterior distribution of θ is proportional to the product of the likelihood and the prior distribution $f(\theta)$

$$f(\theta|x) \propto f(x|\theta)f(\theta) \tag{7.4}$$

The prior distribution $f(\theta)$ should be non-informative, and here we consider it as the product of independent uniform distributions over a sufficiently wide range where values are defined. Once the posterior distribution is specified, we approximate the posterior distribution of the parameters with T random numbers ($t = 1, \ldots, T$) using the MCMC method

$$\theta^{(t)} = (\mu_{\mathrm{b}}^{(t)}, \mu_{\mathrm{a}}^{(t)}, \mu_{\mathrm{h}}^{(t)}, \sigma_{\mathrm{b}}^{2(t)}, \sigma_{\mathrm{a}}^{2(t)}, \sigma_{\mathrm{h}}^{2(t)}, \sigma_{\mathrm{ba}}^{(t)}, \sigma_{\mathrm{ah}}^{(t)}, \sigma_{\mathrm{bh}}^{(t)}) \tag{7.5}$$

I, as Analyst B, did not learn why the posterior distribution can be approximated with MCMC and just memorized it as such. In this case, the joint posterior distribution is a single nine-variable distribution. The joint posterior distribution is a multivariate simultaneous distribution, but it is not generally a multivariate normal distribution.

In the following analysis, we will perform numerous considerations of the posterior distributions of parameters and quantities generated, as well as probability calculations using PHC. However, these are merely areas or mean values of the marginal distributions of a single joint posterior distribution, or distributions of transformed variables. We are simply re-observing the same joint posterior distribution from different angles and perspectives. Therefore, no multiplicity problem arises.

It is entirely acceptable to reanalyze to test hypotheses that occurred to you after seeing the data or even after analysis. The joint posterior distribution exists independently of inspiration or intention and does not change due to them. Reanalysis is simply a matter of looking at the joint posterior distribution from a different viewpoint, and it is not a violation of rules.[4] Rather, it's about diligently searching for unknown treasures of knowledge already embedded in the joint posterior distribution.

7.2.2 *In the Case of a Learner Who Memorized that the Posterior Distribution Can Be Obtained by MCMC Without Understanding*

RQ.1 Evaluate the Effect Size of Weight Loss Not with Eq. (3.9) **but with Eq.** (3.8)
Understood. I will evaluate using both Eqs. (3.8) and (3.9). The posterior distribution of the within-group standard deviation in the denominator of Eq. (3.8), using the average of the variances of the two groups, is approximated by

$$\sigma^{(t)}_{\text{within}} = \sqrt{(\sigma_{\text{b}}^{2(t)} + \sigma_{\text{a}}^{2(t)})/2} \tag{7.6}$$

The posterior distribution of the standard deviation of the weight difference[5] in the denominator of Eq. (3.9) is approximated by

$$\sigma^{(t)}_{\text{difference}} = \sqrt{\sigma_{\text{b}}^{2(t)} + \sigma_{\text{a}}^{2(t)} - 2\sigma^{(t)}_{\text{ba}}} \tag{7.7}$$

Therefore, the posterior distributions of Eqs. (3.8) and (3.9) are approximated by the generated quantities

$$\delta^{(t)}_{\text{within}} = \frac{\mu_{\text{b}}^{(t)} - \mu_{\text{a}}^{(t)}}{\sigma^{(t)}_{\text{within}}}, \quad \delta^{(t)}_{\text{difference}} = \frac{\mu_{\text{b}}^{(t)} - \mu_{\text{a}}^{(t)}}{\sigma^{(t)}_{\text{difference}}} \tag{7.8}$$

These are familiar equations when the subscript (t) is removed. In short, even if one does not know the principles of MCMC at all, as long as one understands the definition formulas, even a student in the humanities can easily derive the posterior distribution of functions of parameters by themselves.

---

[4] In the case of significance testing, conducting subtests not pre-registered later on is a rule violation because the intention changes and the sample distribution changes.

[5] The variance of the weight difference is calculated as follows: $V[x_{\text{b}} - x_{\text{a}}] = E[(x_{\text{b}} - x_{\text{a}} - (\mu_{\text{b}} - \mu_{\text{a}}))^2] = E[((x_{\text{b}} - \mu_{\text{b}}) - (x_{\text{a}} - \mu_{\text{a}}))^2] = E[(x_{\text{b}} - \mu_{\text{b}})^2 + (x_{\text{a}} - \mu_{\text{a}})^2 - 2(x_{\text{b}} - \mu_{\text{b}})(x_{\text{a}} - \mu_{\text{a}})] = \sigma_{\text{b}}^2 + \sigma_{\text{a}}^2 - 2\sigma_{\text{ba}}$.

RQ.2 Evaluate Weight Loss in Measurable Units Positively
The posterior distribution of the average change in weight before and after the diet is approximated by

$$\mu_{\text{difference}}^{(t)} = \mu_{\text{b}}^{(t)} - \mu_{\text{a}}^{(t)} \tag{7.9}$$

Not only the average difference of the group but also, if we want to consider the distribution of individual weight differences, we determine the posterior predictive distribution of the weight difference $x_{\text{difference}}$ as

$$x_{\text{difference}}^{*(t)} = x_{\text{b}}^{*(t)} - x_{\text{a}}^{*(t)} \tag{7.10}$$

The posterior predictive distribution of weight marked with an asterisk (*) is obtained using trivariate normal random numbers as

$$(x_{\text{b}}^{*(t)}, x_{\text{a}}^{*(t)}, x_{\text{h}}^{*(t)}) \sim \text{multi\_normal\_rng}_3(\boldsymbol{\mu}^{(t)}, \boldsymbol{\Sigma}^{(t)}) \tag{7.11}$$

along with the posterior predictive distribution of height $x_{\text{h}}^{*(t)}$.

RQ.3 Evaluate Weight Loss in Terms of Obesity Level
The definition of **obesity level** BMI is

$$\frac{\text{weight in kg}}{(\text{height in m})^2} \tag{7.12}$$

Therefore, the posterior predictive distributions of BMI before and after the diet are approximated respectively by

$$\text{BMI}_{\text{before}}^{(t)} = \frac{x_{\text{b}}^{*(t)}}{(x_{\text{h}}^{*(t)}/100)^2}, \quad \text{BMI}_{\text{after}}^{(t)} = \frac{x_{\text{a}}^{*(t)}}{(x_{\text{h}}^{*(t)}/100)^2} \tag{7.13}$$

The posterior predictive distribution of the difference in BMI before and after the diet is approximated by

$$\text{BMI}_{\text{difference}}^{(t)} = \text{BMI}_{\text{before}}^{(t)} - \text{BMI}_{\text{after}}^{(t)} = \frac{x_{\text{b}}^{*(t)} - x_{\text{a}}^{*(t)}}{(x_{\text{h}}^{*(t)}/100)^2} \tag{7.14}$$

The division by 100 is for converting from cm to m.

Table 7.3 Posterior distributions and predictive distributions of indicators

	EAP	Post.sd	0.05	0.5	0.95
σ_{within}	5.03	0.77	3.93	4.94	6.44
$\sigma_{\text{difference}}$	0.63	0.10	0.49	0.62	0.81
δ_{within}	0.40	0.06	0.30	0.40	0.51
$\delta_{\text{difference}}$	3.23	0.51	2.40	3.22	4.09
$\mu_{\text{difference}}$	1.98	0.12	1.79	1.98	2.17
$x^*_{\text{difference}}$	1.98	0.65	0.92	1.98	3.04
$\text{BMI}_{\text{before}}$	21.1	0.70	19.9	21.1	22.3
$\text{BMI}_{\text{after}}$	20.3	0.80	18.9	20.3	21.6
$\text{BMI}_{\text{difference}}$	0.82	0.29	0.36	0.80	1.31

7.3 PHC Makes the Meaning of Analytical Results More Tangible

The summary statistics of the posterior distributions and predictive distributions for the indicators introduced in the previous section are shown in Table 7.3.

RQ.1 Evaluate the Effect Size of Weight Loss Not with Eq. (3.9) **but with Eq.** (3.8)
The EAP estimate of the variability of weight within the group, σ_{within}, is 5.03 kg. The EAP estimate of the average weight loss, $\mu_{\text{difference}}$, is 1.98 kg. With PHC $(0.30 < \delta_{\text{within}}) = 0.95$, we can assert with 95% credibility that "the average weight loss is more than 30% of the group's weight variability." Using a Bayesian approach, insights usually obtained only from independent two-group experimental data can easily be obtained from corresponding two-group experimental data.

The EAP estimate of the variability of weight loss, $\sigma_{\text{difference}}$, is 0.63 kg. With PHC $(2.40 < \delta_{\text{difference}}) = 0.95$, we can assert with 95% credibility that "the average weight loss is more than 2.4 times the variability of weight loss."

0.3 times and 2.4 times are entirely different multipliers. Using a Bayesian approach, two completely different insights can be obtained from a single set of data, clarifying the difference between insights from between-subject and within-subject sources. In contrast, with significance testing, only specific insights can be obtained from either a corresponding two-group or an independent two-group experimental design. This leads to a situation[6] where learners find it difficult to recognize that the insights provided by both are entirely different.

RQ.2 Evaluate Weight Loss in Measurable Units Positively
With PHC $(1.79 < \mu_{\text{difference}}) = 0.95$, we can confidently state that "the average weight loss with this diet method is more than 1.79 kg" with 95% certainty. How-

[6] Since they are different experimental designs intended to yield different insights, they should not be comparable. However, learners often memorize ambiguous content such as "it's easier to obtain significant differences with corresponding two-group differences than with independent two-group differences." This misunderstanding arises from the overemphasis on obtaining significant differences.

ever, this statement pertains to the population parameter $\mu_{\text{difference}}$, not to individual participants. For individuals, since PHC $(0.92 < x^*_{\text{difference}}) = 0.95$, we can say with 95% credibility that "an individual participating in this diet can expect to lose more than 0.92 kg."

It is also possible to calculate the PHC for the hypothesis "this diet method has no effect." For instance, let's consider a Region of Practical Equivalence (ROPE) of "average weight change after one month being $\pm$ 1 kg."[7] In this case, PHC $(-1.0 < \mu_{\text{difference}} < +1.0)$ was found to be 0.000. Under this premise, we can declare that the probability that the hypothesis "this diet method has no effect" is correct is 0.0%. Similar calculations can be made for other indicators regarding the probability that the "hypothesis of being within the ROPE" is correct. It is essential to use domain knowledge to non-statistically evaluate the reference points for ROPE.

RQ.3 Evaluate Weight Loss in Terms of Obesity Level

The probabilities were PHC $(0.36 < \text{BMI}_{\text{difference}}) = 0.95$, PHC $(0.80 < \text{BMI}_{\text{difference}}) = 0.50$, and PHC $(1.31 < \text{BMI}_{\text{difference}}) = 0.05$. Participating in this diet method implies that the probability of a reduction in BMI by more than 0.36 is 95%, the probability of a reduction in BMI by more than 0.80 is 50%, and the probability of a reduction in BMI by more than 1.31 is 5%.

Obesity level BMI is not a statistical indicator like mean or standard deviation but a medical indicator. There are countless indicators in substantive scientific fields that are not directly related to statistics.

Adopting a Bayesian approach allows us to approximate posterior and predictive distributions from definition formulas in many cases. Most learners of statistical analysis do not become statisticians. In their active roles in substantive scientific fields, they can surpass their teachers by devising posterior and predictive distributions of indicators unknown to statisticians. This ability may well represent a true partnership between statistics and substantive science. However, this dream is unattainable with significance testing, where individual ingenuity is not feasible.

7.4 Properties of PHC Curves

Various PHC curves are shown in Fig. 7.1.

Figure a represents the PHC curve of the posterior distribution of δ_{within}, the mean difference standardized by the within-group standard deviation σ_{within}. As a larger value is desirable for this indicator, PHC $(c < \delta_{\text{within}})$ is plotted, resulting in a monotonically decreasing function.

Figure b is the PHC curve of the posterior distribution of the standard deviation of weight difference, $\sigma_{\text{difference}}$. A diet method is preferable when the effects are stable. With PHC $(\sigma_{\text{difference}} < 0.81) = 0.95$, we can say with 95% credibility that

[7] A 1 kg weight loss is a target mentioned by doctors, so it's not originally a ROPE. By setting the ROPE generously in a direction unfavorable to the analyst, it turns out that the probability that the hypothesis "this diet method has no effect" is correct is zero.

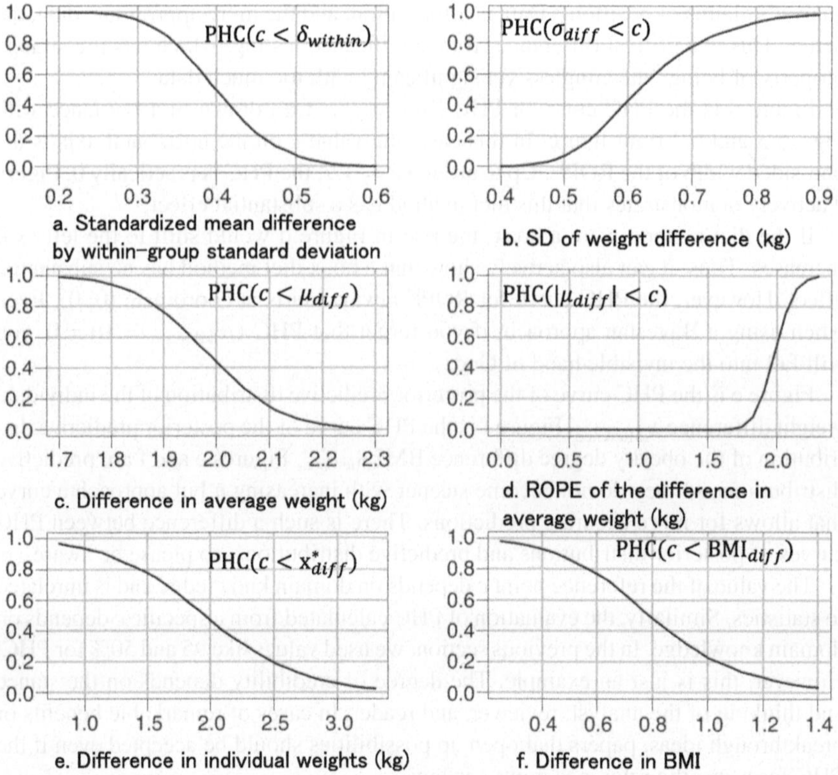

Fig. 7.1 Various PHC curves

"the variability in weight loss is less than 0.81 kg." As a smaller value is desirable, PHC ($\sigma_{\text{difference}} < c$) is plotted, creating a monotonically increasing curve.

Figure c shows the PHC curve of the posterior distribution of the average weight difference $\mu_{\text{difference}}$. It is a monotonically decreasing function as a larger value is desirable. Figures a, b, and c are PHC curves of posterior distributions, so they change from a gentle to a steep shape with the increase in n. Let's consider the meaning of this using Figure c.

When researchers spare their efforts and n is small, the PHC of a gently sloping decreasing function cannot be significant in the region where the value c on the horizontal axis is large. This means that the risk of an analyst's erroneous claim being accepted due to the instability caused by a small n is extremely low. In other words, when n is small, it is difficult for an analyst's claim to be accepted. As the proof of the value of research principally lies on the analyst's side, this characteristic is highly rational.

As n increases and the curve shape becomes steeper, the PHC at any c approaches 0 or 1.[8] Therefore, with a large n, clear judgments can be made, irrespective of the

[8] Of course, the dividing point does not necessarily give PHC = 0.5 at the current $n = 30$.

analyst's claims or position.[9] Data is information, and the more appropriate data, the better. This is a critical difference from significance testing, which has the strange property of being "meaningless yet significant" with too much data.

Figure d is the PHC curve of PHC ($|\mu_{\text{difference}}| < c$), drawn in accordance with Figs. 5.3 and 6.2 right figure. In this case, the value c on the horizontal axis is the one-sided width of the ROPE. Up to around $c = 1.7$, the PHC is practically 0. Figure d actively demonstrates that this diet method has a substantial effect.

If the diet method were a hoax, the rise in Figure d would shift to the left as n increases. Thus, it can also actively show that a hoax diet method has no substantial effect. However, the PHC curve for ROPE always starts at coordinate (0, 0). Even when using a Bayesian approach, if you forget that PHC ($\mu_{\text{difference}} = 0$) = 0, you will fall into the invisible hand of God.

Figure e is the PHC curve of the posterior predictive distribution of the individual weight difference $x^*_{\text{difference}}$. Figure f is the PHC curve of the posterior predictive distribution of the obesity degree difference $\text{BMI}_{\text{difference}}$. Figures e and f are predictive distributions, so they do not become steeper with increasing n but approach a curve that allows for more accurate predictions. There is such a difference between PHC curves of posterior distributions and predictive distributions, so please be aware.

The value of the reference point c depends on domain knowledge and is unrelated to statistics. Similarly, the evaluation of PHC calculated from a specific c depends on domain knowledge. In the previous section, we used values like 95 and 50% for PHC. However, this is just an example. The degree of credibility depends on the stance and thinking of the analyst, reviewer, and reader. In cases of remarkable benefits or breakthrough ideas, papers that open up possibilities should be accepted even if the PHC for a specific reference point c is low.

Unlike significance testing, there is no need to set a reference point, conduct power analysis, and predetermine n. You will fall into the invisible hand of God if you consider the reference point as zero (or no effect).

However, before or after collecting data, the analyst does not need to specify the value of the reference point c. It suffices to show the PHC curve that aggregates and visualizes the relationship between the reference point c and PHC. By observing the PHC curve, a comprehensive evaluation of the value of the indicator on the horizontal axis is possible.

7.5 The Future of Statistical Education

When teaching statistical courses to students in the humanities and social sciences, it becomes essential in the learning sequence to omit some parts of mathematical proofs. It is impossible to avoid top-down approaches. A top-down approach means that students memorize the content without mathematical proofs or derivations of formulas.

---

[9] Please review Fig. 5.8.

7.5.1 Top-Down Approach and Learned Helplessness

In introductory education on significance testing, test statistics following normal, t-, F-, and χ^2-distributions are introduced. These test statistics cannot be derived without knowledge of university-level analysis. Therefore, it becomes necessary for students in the humanities and social sciences to memorize these test statistics in a top-down manner. This unavoidable top-down approach to test statistics is a form of simple memorization that does not apply to other significance tests. As it is not applicable elsewhere, unlearned tests that become necessary later must be looked up in books and memorized again. The top-down approach to test statistics ultimately hinders autonomous thinking.

The following is a famous experiment related to learned helplessness.[10]

Seligman's Dog Experiment
An experiment is conducted where dogs are restrained in an inescapable condition and repeatedly subjected to strong electric shocks on their limbs. Initially, the dogs struggle desperately to escape. However, gradually, they give up and continue to endure the pain passively.

Next, the dogs are released from their restraints and placed in a room divided by a low fence into two sections. In this room, an electric current flows only through the floor of the section where the dog is, after a warning light is illuminated. The dogs could easily escape to the other section by jumping over the fence. However, they no longer react to the warning light at all, enduring the pain and continuing to receive electric shocks. This is because they have learned a sense of helplessness itself.

Mr. A, who appeared in this chapter, is like Seligman's dog. Even when told that a comparison with others is necessary, he only attempts comparisons using Eq. (3.9). This is because he has been educated to simply memorize test statistics. When repeatedly presented with the notion, "You can only memorize values like z-, t-, F-, and χ^2-values," students in the humanities and social sciences fall into learned helplessness. They learn a sense of uncontrollability and powerlessness. They end up with the impression that "statistics is just a tool for data analysis, a subject of memorizing tedious procedures." They never even consider devising their own analytical methods, ultimately leading to the development of students who are like Seligman's dogs. It is unfortunate because statistics is an exciting and stimulating field with plenty of room for creativity using generated quantities and likelihoods.

[10] Seligman, M. E. P., Maier, S. F., & Solomon, R. L. (1971). Unpredictable and uncontrollable aversive events. In F. R. Brush (Ed.), *Aversive conditioning and learning*. New York: Academic Press, pp. 347–400.

7.5.2 Top-Down Approach and Sense of Control/Efficacy

A top-down approach is not necessarily a shortcut or a compromise. For students in the humanities and social sciences, simplifying the path can often be educationally appropriate. There are indeed crucial subjects where knowing only the outcomes accurately as "that's just how it is" is important.

In the elementary learning process of Bayesian statistics, the most mathematically advanced topic is MCMC. In elementary statistical education for students in the humanities, it is often the case that MCMC is memorized in a top-down manner as "a method for approximating the posterior distribution with random numbers." However, the principle of MCMC itself is not necessary[11] to conduct data analysis or model construction. This is because the mathematical principle of MCMC is irrelevant for devising ways to express phenomena through likelihood or determining posterior distributions of indicators not taught by the teacher.

Students who understand the posterior distribution of the generated quantity of the mean difference can also determine the posterior distribution of the coefficient of variation (σ/μ) if necessary, even without being taught. They can find distributions for indicators that only appear in specialized research as long as there is a definition formula. For example, in the previous case, we derived the distribution of obesity level BMI. The property of "opening possibilities for inferential statistical consideration if the definition formula is known" is a notable strength of the Bayesian approach. Applying this property will dramatically advance analysis in various substantive scientific fields.

Such autonomous growth is entirely unachievable with significance testing. The greatest educational merit of modeling with likelihood is not just memorizing fixed procedures, but learning the sense of control and efficacy that "it is possible to devise analytical methods by oneself."

7.5.3 Fostering a Sense of Competence and Control in Modeling the Data Generation Process

The future of statistical education lies in fostering knowledge, skills, and interest to answer the fresh question, "How was this intriguing phenomenon generated and appeared as data before me?" using likelihood. In other words, the goal is to help

[11] For intellectually curious undergraduates with knowledge of Calculus II, the essence of MCMC can be taught. See Chaps. 4 and 5 of [15]. Undergraduates can understand the essence of MCMC. If comparing the mathematical level of MCMC to the height of a mountain, it's like a small hill climbed after lunch. On the other hand, deriving test statistics is significantly more complex. It is not an overstatement to say that almost no statistician, including the author, can derive sample distributions effortlessly. The mathematical level here is equivalent to Mount Everest. Since free derivation as required for analysis is impossible, psychology students unavoidably become like Seligman's dogs.

students acquire the ability to model the data generation process using probability distributions. This is, in fact, an incredibly fun and exhilarating intellectual activity, akin to solving a puzzle.

However, many students who start with significance testing fail to acquire the analysis paradigm of considering phenomena through likelihood, and thus complete their statistical education without this essential skill. This is because elementary statistical education does not have an abundance of time or credits. The content taught must be carefully selected. Education on significance testing should be limited to the level necessary for reading past papers, and without deviation, educators should focus on teaching "thinking about the data generation process using likelihood" from the beginning. Education should be based on the likelihood principle.

Chapter 8
Tampering and Cover-Up

The Expanding Darkness of a Black Cave.

The issues raised in Part I have been largely resolved in Part II. However, these solutions presume the absence of malice on the part of analysts and contributors. In other words, until now, this discussion has revolved around the difficulties faced by conscientious analysts and their solutions. Yet, the reproducibility of research is significantly hindered by malicious intentions. This chapter, being the final one, delves into the relationship between malicious actions and statistical conclusions.

8.1 Tampering

The greatest threat to the reproducibility of research is **data tampering**. Data tampering involves altering data to different values from those obtained. For example, the true sorted data of "Weight Difference (kg)" in the diet experiment data (Fig. 2.1 and Table 2.1) were

```
- 4.4, - 2.5, - 1.9, - 1.1, - 1.1, - 0.8,  1.0,  1.1,  2.6,  2.7,
  2.8,   4.7,   4.7,   4.8,   5.5,   6.3,  7.1,  7.3,  7.6,  8.4.
```

The first six individuals had gained weight during the diet, which is inconvenient data for proving the effectiveness of a diet method. Therefore, tampering, such as removing the negative sign from the data of these six individuals to falsely indicate weight loss, is an act of data tampering.

Similarly, altering data to show unhealed patients as fully recovered is also tampering. Most cases of tampering likely involve malice. Sadly, acts of tampering are not infrequently reported. They sometimes come to light through internal whistle-

H. Toyoda, *Statistical Significance and the PHC Curve*,
https://doi.org/10.1007/978-981-97-7748-8_8

blowing, disclosure of experimental notebooks, or replication experiments. However, statistical methods are fundamentally powerless against tampering. Neither significance testing nor calculating the probability that a hypothesis is correct (PHC) contributes much to preventing or detecting tampering. Moreover, if there is malice, tampering is easy to execute.

It's embarrassing to raise a white flag from the start, but statistics are almost helpless once data has been tampered with. Regardless of the statistical analysis conducted, tampering with data is strictly prohibited for anyone involved in data analysis. It is an absolute violation.

8.2 Cover-Up

Another threat to the reproducibility of research is **data cover-up**. Data cover-up is the act of not reporting a portion of the observed data.[1] Many cases of cover-up are also believed to involve malice.

For example, regarding the previous "weight difference (kg)" data, we think that the reason they gained weight despite being on a diet is because they did not participate in the experiment seriously. The act of excluding from the analysis six persons with negative values as **outliers** by later giving reasons such as, "This method targets serious participants," is a cover-up of data.

Tampering is an active act committed against one's conscience, accompanied by guilt. In contrast, cover-up is a passive act of "pretending not to see," thereby reducing feelings of guilt. This makes cover-up particularly pernicious. Moreover, as in the following examples, cover-up can occur unconsciously over multiple variables and periods.

8.2.1 Cover-Up by Thinning Out Variables

Measure numerous criterion variables from two groups of subjects. For instance, when comparing a group that practiced a health method with a control group, measure variables such as blood pressure, weight, body fat percentage, feelings of refreshment, depressive feelings, and so on, up to the number of times they won a sweets lottery or saw pleasant dreams during the experiment period. Analyze each variable for differences, interpret those with differences, and report only them. The rest are treated as if they never existed. This is data cover-up by thinning out variables.

PHC (research hypothesis) does not need to be corrected for any number of calculations. Unlike significance testing, it is free from the problem of multiplicity. Not needing adjustment is convenient, but computing an enormous number of PHCs will

[1] 逆に分析者の主張に沿ったデータを選ぶ行為を**チェリーピッキング**といいます。またそのデータを**チャンピョンデータ**と呼ぶこともあります。

result in some inconsistency with the correct knowledge in the relevant field. It is possible that the experimental group happened to win more ice cream sticks in the lottery during the period in question. You might happen to have a lot of pleasant dreams.

However, if all data are honestly disclosed, there is no worry. The intuition that "with so many variables measured and probabilistic statements made, a certain proportion of inconsistencies will naturally occur" is valid. Probabilistic statements are inherently like that.

The simultaneous posterior distribution calculated from the simultaneous likelihood of multiple variables must be reported without alteration. Thinning out variables results in showing a posterior distribution from a likelihood different from the actual data generation mechanism. This disables the principle of likelihood. The abovementioned natural intuition doesn't apply, and even PHC cannot address data cover-up.

8.2.2 *Cover-Up by Selecting Periods*

Suppose no substantial effect is observed in an experiment similar to Table 2.1 on dieting. However, if similar experiments are persistently repeated every month, data showing an effect might eventually be observed. In this case, only the data from the last experiment are reported. The rest of the experiment will not be reported as if it never happened in the first place. This makes it look like an effect that is not really there. This is data cover-up by selecting periods. It involves picking a convenient part from a long series of data and presenting it as if it were correctly and independently observed. In this case, the likelihood does not represent the reality of the data generation, and the correct PHC (research hypothesis) is not calculated.

Cover-up by selecting periods is a more profound problem compared to cover-up, which selecting variables. Authors of papers are not fundamentally obligated to describe what experiments they have previously conducted. There is no obligation to report failed experiments in the peer-review culture. This makes it easier to avoid guilt.

If each experiment was conducted with slight variations in conditions, the problem becomes even more severe. In this case, they are effectively different experiments. Then, the experimenter feels no guilt at all. Even trivial changes in experimental conditions, like changing the color of T-shirts worn by subjects in the experimental group, can make each seem like a different experiment if the experimenter believes they are different. But this certainly impairs the reproducibility of experiments.

8.3 HARKing and Cover-Up

Formulating research hypotheses after viewing the data is known as **HARKing** (Hypothesizing After the Results are Known). Reading the spelling, HARKing itself does not imply a negative connotation. However, in significance testing, regardless of whether there is tampering or cover-up, HARKing is a violation of the rules. One must not add tests to verify hypotheses that seem successful after viewing the data. The original sample distribution changes and such tests become zombie tests indistinguishable from "intention 1 human" on the surface. It also leads to multiplicity problems. Therefore, the original p-values are not reported.

When using PHC, some HARKings are permissible, and those that are not. Let's illustrate recommended HARKing with the diet data from Chap. 7. For the sake of explanation, let's assume that the data in Table 7.1 was initially collected to evaluate the effect of a beauty diet. Beauty has a relative comparison aspect. Therefore, following the original plan, the effect of the diet was evaluated δ_{within} in Eq. (3.8). However, only modest credibility could be obtained with PHC $(0.3 < \delta_{\text{within}}) = 0.95$. Ultimately, it was judged that the method was insufficient as a beauty treatment. If left at that, the experiment would be deemed a failure.

In order to make the most of the data, we tried to HARKing to find the value in terms of weight loss stability. Suppose that in this process, after looking at the data, we re-evaluate the effect by the $\delta_{\text{difference}}$ in Eq. (3.9). This led to the discovery of a significant effect PHC $(2.40 < \delta_{\text{difference}}) = 0.95$. Therefore, he changed his original research policy and rewrote his thesis as a diet for the prevention of lifestyle-related diseases. This is an example of recommended HARKing.

The simultaneous posterior distribution, unaltered by tampering or cover-up, is independent of whether the research hypothesis was conceived beforehand. Therefore, PHC is not affected by whether the hypothesis was thought of after the fact. In short, up to Chap. 7, calculating PHC through HARKing is a legitimate research method. Confessing to HARKing in a paper is an honest act, but whether or not one confesses does not influence probability assessment.

However, when combined with tampering or cover-up, the simultaneous posterior distribution changes. Reporting PHC (research hypothesis) that suits the analyst's preference becomes a violation of the rules. Not limited to HARKing, when using PHC, one should focus on whether their actions are altering the simultaneous posterior distribution. This helps distinguish what is permissible and what is not.

8.4 Rethinking the Pre-registration System

8.4.1 Review of Pre-registration in the Absence of Cover-Up

Analysts conducting significance testing needed to pre-determine n through power analysis before data collection. Data collection should not be stopped midway just because the p-value falls below 5%. Nor should data be further collected if the p-value calculated with the planned n exceeds 5%. Properly executing significance testing was quite challenging.

There was no physical evidence of rule violations. It is impossible to tell from the final analysis whether they are legitimate "intent 1 humans" executed with a predetermined n or zombies based on some other intention. Even if the raw data is disclosed, the distinction is impossible. Therefore, to claim "this is not a zombie test," one needs to pre-register the n based on evidence. Even if the p-value is peeked at, the knowledge cannot be utilized if n is fixed. It's a requirement based on mathematical properties and is necessary not only for peer review but also for theses and conference presentations. Without a certificate of pre-registration, significance testing cannot be trusted in any application. Therefore, the fact of pre-registration needs to be permanently guaranteed. This is an irony. Don't Say 'Statistically Significant'—a cumbersome procedure for everyday use should be prohibited in significance testing, which is flawed.

In contrast, there is no need to pre-determine n before data collection when using PHC. n can be freely decided while observing the situation. If n is small, the analyst, responsible for demonstrating the value of the research, finds it harder to make their case. If n is large, the truth is easily revealed, regardless of the researcher's position. Therefore, up to Chap. 7, pre-registration of research is not necessary when using PHC.

8.4.2 Differences Between p-Values and PHC in Pre-registration

However, as learned in the previous section, even when using PHC, cover-up threatens research reproducibility. To prevent the ill effects of cover-up, it is effective to pre-register "detailed information on variables, levels, attributes, groups, measurement methods, analysis models, and other detailed procedures of experiments and surveys." Pre-registering these details with a third party can prevent much cover-up. Once registered, one cannot claim "it wasn't done." Even when using PHC, pre-registration is effective in ensuring research reproducibility.

In order to conduct significance testing, it was necessary to pre-register n on the basis of evidence, even in the absence of concealment or other malicious intent. The pre-review of significance testing requires comprehensive knowledge of power analysis, multiple comparisons, and sequential testing. Therefore, it is very challenging for academic journals in fields that use statistics as a tool (such as medicine and psychology) to have reviewers fully equipped with this knowledge. This is also an irony. Such reviewers cannot be prepared. Therefore, the pre-registration system cannot prevent the ill effects of significance testing. Don't Say 'Statistically Significant'—to stop unreproducible papers, significance testing should be prohibited.

However, PHC does not require pre-registering n in advance. There is no need for knowledge of power analysis, multiple comparisons, or sequential testing. Just pre-registering research procedure information allows one to escape the threat of cover-up. In other words, mere substantial scientific knowledge (such as medical or psychological knowledge) of the journal suffices for pre-review. Therefore, using

PHC makes it easier to procure reviewers and smoothly operate the pre-registration system.

8.4.3 Publication Bias and Peer-Reviewed Pre-registration System

As previously mentioned, research pre-registration involves a certain level of review. Further, a system exists where "papers that have undergone peer review and passed pre-registration are guaranteed publication." The advantage of this peer-reviewed pre-registration system that assures publication lies in its ability to resolve **publication bias**. For instance, consider independent studies conducted worldwide to verify the efficacy of a treatment for a particular disease. Due to individual differences, experimental conditions, and probabilistic variations, studies may show varying degrees of efficacy, from noticeable effects to none at all. Studies that observe an effect are more likely to be published, while those that do not are often deemed failures and less likely to be published, depending on the degree. This leads to a perception within the academic community that the treatment is more effective than it actually is, known as publication bias or **the file drawer problem**.

The peer-reviewed pre-registration system with guaranteed publication also has the advantage of preventing data tampering and concealment. If all results are to be published regardless of their nature, there is no need to take risks associated with data tampering or concealment, thus eliminating any motive for malice. However, this system functions optimally only in cases where the effects of well-known themes within the academic community, such as estimating the efficacy of established treatments, are being evaluated. It is limited to studies where a lack of effect also carries academic significance.

For example, in research to discover entirely new cancer drugs, substances are administered to mice, and the results are observed. Of course, substances are chosen not randomly but based on prior research, selecting a group of promising substances with a particular molecular structure. The motivation, hypothesis, and experimental design are perfect, and suitable for peer-reviewed pre-registration. However, out of thousands of substances, perhaps only one might prove to be an effective cancer drug. Publishing experiments on ineffective substances would worsen SN ratio (Signal-to-Noise ratio) of academic journals.

In many fields, researchers striving to discover original phenomena, even with perfect motivation, hypothesis, and experimental design, often struggle to achieve meaningful results, akin to the discovery of cancer drugs. If failed experiments are consistently published in academic journals, it leads to a flood of irrelevant papers. Thus, a peer-reviewed pre-registration system with a commitment to publication works only to a limited extent.

8.4.4 *PHC and Pre-registration*

In accordance with the decision of the American Statistical Association, the use of significance testing should be urgently prohibited. With that as a premise, let's discuss the relationship between PHC and pre-registration. Pre-registration of detailed procedures in experiments and surveys effectively addresses data concealment. However, making pre-registration a necessary condition for paper acceptance is not always appropriate. As soon as it becomes a requirement, it fosters malice to circumvent it. If malice is assumed, any form of pre-registration can be manipulated by tampering with the data itself. Against such malice, neither the pre-registration system nor any statistical analysis is effective. Therefore, enforced pre-registration is not a panacea for solving reproducibility issues. Like transactions in economic activities that operate under the concept of trust, research activities should also fundamentally be based on trust, as the peer-review system does not balance out in terms of social costs otherwise.

The human heart has its weaknesses. The tendency to ignore inconvenient data and inadvertently conceal it is one such weakness. When embarking on a diet, some people physically distance themselves from sweets to overcome their weaker selves. Similarly, voluntarily registering research plans and progress can be effective in resisting the temptation to conceal. Sites for voluntary pre-registration already exist, and it's also feasible to share experimental notes from one's personal website. When pre-registration is a voluntary act of self-discipline, it can have a significant impact on oneself and others. However, this increases the risk of ideas being stolen by peers or reviewers.

Research is an activity that seldom yields rewards. To make a discovery, one must trust their intuition, change conditions, and subjects, and struggle adaptively and exploratively. For resourceful research activities, pre-registration should not exist. If discoveries are missed or ideas are stolen due to the costs of pre-registration, it defeats the purpose. Without tampering or concealment, PHC does not require pre-registration, unlike significance testing. The most important thing, needless to say, is for you to quickly discover phenomena that are useful, surprising, and impactful. In such times, PHC, which does not require pre-registration, will contribute to your dynamic research activities as a good person.

8.5 Saving Dying Statistics!

Contrary to statistics, the field of AI (Artificial Intelligence) aimed for sufficient conditions for scientific progress from its inception. For example, Go AI does not reject hypotheses like "placing stones randomly"; instead, it aims for sufficient conditions for academic progress, such as winning against first-dan players. AI for future prediction (of stock price fluctuations) is not concerned with rejecting a "prediction accuracy rate of 50." This is natural. Where does this difference between statistics and AI come from?

Gray geese assume that the first human they see after hatching is the parent bird and follow it. It is said that it is difficult to correct the situation after that. In animal behavior, this is called imprinting. Early experiences are crucially important. Current statistics education imprints the worst idea that "research is sufficient with verifying necessary conditions."

"The mistaken research paradigm of not aiming for sufficient conditions," imprinted by introductory materials, remains uncorrected for many researchers throughout their lives, like gray geese. If this continues, statistics will be abandoned by science as a useless tool and be driven out by AI research. Dying statistics will indeed die. This mystery novel is non-fiction, ongoing, and unfinished. A curriculum shift away from significance testing is urgently needed.

Q & A

Does the Author Deny Frequentist Statistics?

No, neither the author nor this book denies **Frequentist statistics**. So far, I have consistently conducted research based on the likelihood principle, focusing on **Covariance Structure Analysis (Structural Equation Modeling)** and **Item Response Theory**. I am neither a **frequentist** nor a **Bayesian**; I use both approaches. The root of the nuisance that statistics is currently causing in neighboring sciences lies in significance testing, which is only one aspect of frequentist theory. Significance testing has already fulfilled its historical mission. As suggested in the journal Nature, I believe that it is time to "abandon the entire concept of statistical significance."

Prior Distributions Are Subjectively Chosen and Therefore Not Trustworthy

It is true that prior distributions are subjectively chosen. However, the distributions that constitute the likelihood (such as the binomial or normal distribution) are also subjectively selected. The data distributions used in significance testing are subjectively chosen as well. In data analysis, the process of subjectively choosing a distribution based on the analyst's judgment is essential.

Why Is the Bayes Factor Not Recommended in This Book?

Let's use the inference of the difference in means between two independent groups as an example to explain the reasons.

(1) The current mainstream **Bayes Factor (BF)** compares models where the means of two groups are equal (M_0) and not equal (M_1), similar to the t test. M_1 represents a linear hypothesis, while M_0 represents a point hypothesis. M_0 is almost surely (a.s.) covered probabilistically by M_1. Since points and lines are of different dimensions, a comparison based on substantive science is not possible. This means that when n is small, M_0 is favored, but as n increases, even if it's scientifically meaningless, M_1 tends to be favored. It's like the p-value. Just changing the cutting point leads to the invisible hand of God, similar to significance testing.

H. Toyoda, *Statistical Significance and the PHC Curve*,
https://doi.org/10.1007/978-981-97-7748-8

(2) One of the drawbacks of significance testing was that the p value cannot be substantively scientifically interpreted. The value of BF also cannot be substantively scientifically interpreted. It is impossible to decide from a substantive scientific context how many times better M_1 should be than M_0 to adopt M_1. Eventually, we have to set an arbitrary number equivalent to $p < 0.05$, repeating the same mistake as with p-values.

(3) BF is highly sensitive to the parameters of the prior distribution. For example, it is sensitive to values such as the scale parameter of the Cauchy or t distribution, or the range of a uniform distribution. Yet, there is no substantive scientific basis for determining the values of these parameters of the prior distribution. For instance, if the JASP software becomes popular, the scale parameter value chosen as a default could become the second 0.05.

Those who wish for a difference or those who do not may set values convenient for their purposes. It's a natural consequence of human nature. Therefore, irrelevant to domain knowledge, debates start during peer review. The arbitrariness of $p < 0.05$ was a drawback of significance testing. On the other hand, with BF, there are two directions of arbitrariness not based on substantive science: the cutting point and the parameters of the prior distribution, making the situation even worse.

By the way, if a uniform distribution is adopted as the prior distribution and it sufficiently covers the likelihood, the value of PHC hardly changes even if the interval is changed. The value of BF, however, changes significantly.

Why Don't You Use the Term p-Hacking?

***p*-Hacking** refers to a situation where, despite not being academically useful, a result of $p < 0.05$ is obtained. p-Hacking arises due to the invisible hand of God, the zombie problem, and the multiplicity problem, all of which can be prevented by using PHC. However, p-Hacking can also occur due to falsification and concealment, which cannot be prevented even with PHC. High PHC calculated from **champion data** obtained through concealment does not guarantee replicable results. Using the term p-Hacking, which is a general term for a state, makes it difficult to discuss and differentiate causes that can be prevented from those that cannot. This book considers each cause of occurrence separately.

What Should Be Kept in Mind When Using PHC?

1. When interpreting PHC, never forget that it represents the probability of a hypothesis being correct, under the acceptance of certain mathematical premises such as "assuming a normal distribution for likelihood and a uniform distribution for the prior." If the premises are not valid, the interpretation derived from PHC becomes correspondingly invalid. For instance, in complex models like hierarchical modeling, it's crucial to question whether the model fits the data. Those interested in handling complex models should definitely learn about **posterior predictive checks** and **cross-validation** from advanced textbooks. However, within the scope of this book and the literature [8, 17], the models are simple, so there's no need to be overly anxious. Remember, PHC, like p-values and other statistical measures, has mathematical premises.

2. Similar to the null hypothesis, using a completely research-insignificant state (like $\pi = 0.5$) as a reference point will lead you into the invisible hand of God. The same applies when interpreting the PHC curve. Always keep in mind a reference point that meets the sufficient conditions for scientific progress.
3. PHC calculated from falsified data, champion data involving concealment, or data that misrepresent the data generation process cannot be trusted. PHC that alters the simultaneous posterior distribution is not reliable. Please reread Chap. 8.

I Want to Decide *n* in Advance Because Power Analysis Is Difficult

When supervising undergraduate theses, it's impossible for psychology professors to instruct on power analysis for dozens of different tests. If you let students decide *n* in advance, guarantee it forever, and base your testing on that, you won't turn into a zombie. To remain a "single-intention human," **power analysis** is not essential. However, you may face complaints from students saying, "I was forced to decide *n* in advance by the professor, but in retrospect, the power was insufficient, making my work futile (it was doomed from the start)," or "I was made to decide *n* in advance by the professor, but in retrospect, although it was significant, it was psychologically meaningless."

By the way, if you decide in advance on $n = 100$, but only 80 data points are collected, would you make the student repeat the year for that reason? Insisting on significance testing and deciding *n* in advance implies this. In such cases, please tell them, "It's okay to change *n* along the way, there's a method called PHC."

I'm Done When You Say Probability Distribution

Both PHC and *p*-values inevitably require understanding probability distributions due to the nature of the subject. To teach significance testing up to ANOVA and cross-tabulation, you need to teach normal distribution, binomial distribution, multinomial distribution, chi-square distribution, *t* distribution, and *F* distribution. The curriculum advocated in this book only requires normal distribution, binomial distribution, multinomial distribution, and uniform distribution. In fact, the hurdle of distributions required in the teaching process is lower with PHC.

p-Values and PHC May Coincide in Some Cases

It is rare for *p*-values and PHC to coincide or have a simple functional relationship. Generally, they do not coincide or relate in a simple way. Likewise, confidence intervals and credible intervals only coincide in rare cases.

It is known that *p*-values and PHC may coincide or have a simple functional relationship in certain cases, such as when the likelihood is composed of a normal distribution and a sufficiently broad uniform distribution is chosen as the prior distribution, depending on the intention or research hypothesis. However, while *p*-values often lead to misinterpretations, PHC can be intuitively interpreted. Therefore, even in rare cases where they show similar values, PHC should be calculated.

Cases where they do not coincide or have a simple relationship include the following:

1. When considering data generation processes with distributions other than the normal distribution. In **non-normal models**, comparing the location of distributions between two groups can be challenging, especially in obtaining the sample distribution or p-value of the difference in location parameters. Cases where a test method has been proposed are rather rare. On the other hand, PHC allows for relatively easy calculation of the probability of a difference, even in non-normal models.
2. When considering distributions other than a uniform distribution as the prior distribution. For instance, when the properties of the data or previous research narrow down the range of parameters (such as weights or times not being negative), or when using a prior distribution other than a uniform distribution, p-values and PHC do not necessarily coincide or have a simple relationship.
3. When interested in generative quantities. Even when considering the data generation process with a normal distribution, it becomes difficult to obtain p-values when interested in generative quantities, such as the difference in BMI (Body Mass Index) calculated from weight and height between groups. PHC allows for relatively easy calculation of probabilities according to the hypothesis.
4. When interested in the difference in means and there are many groups. In multiple comparisons of significance testing, there are many methods, including Holm's, Hochberg's, Hommel's, Bonferroni's, and Tukey's. This means that for one dataset, significance testing can calculate many different p-values. In contrast, once the model and data are specified, PHC is uniquely determined. It's also easier to calculate the more practical **probability that a conjunctive proposition is correct**. Obtaining the p value equivalent to the "probability that a conjunctive proposition is correct" in significance testing is a challenging task.

Why Can One Live Off Day Trading with a 53% Hit Rate?

Setting non-statistical reference points is very important. In Bem's example, the rate of correctly predicting the location of nude photos was 0.531 [0.506,0.556]. Predicting nude photos is academically meaningless. However, the same parent ratio can be meaningful in predicting the future. In literature [6], I reported a model for predicting binary stock price movements using (not deep learning but) neural networks with backpropagation, automatically collecting information from the internet. At that time, there were few individual investors predicting using scraping or AI, so the investment performance was good.

Here, we discuss a prediction system (sell or buy decision) in a situation where, after purchase, if the stock price moves up or down by 1 yen, the position is closed regardless of profit or loss. A trader with a 10 million yen fund buying 100,000 shares at 100 yen each will either gain 100,000 yen (equivalent to seeing a nude photo) or lose 100,000 yen (nothing). If each transaction's gain or loss is independent, the total gain or loss per day is

$$(c \times 100{,}000 \text{ yen} + (1-c) \times (-100{,}000 \text{ yen})) \times \text{number of trades}$$

Let's assume the probability of predicting stock price movement is the same 0.531 [0.506, 0.556] as in Bem (2011). If the full fund is utilized for 10 trades per day, the daily profit or loss becomes 62,000 yen [12,000 yen, 112,000 yen], sufficient for living. However, with a 1 million yen fund and then-prevailing commissions, this parent ratio was not sufficient for living. The same hit rate has value for future prediction as a trading system for a trader with 10 million yen, unlike nude photos.

Won't Some Academic Journals Go Out of Business if PHC Is Used?

I once received the following confession from a researcher:

"When I hear 'there is a statistically significant difference', I feel like it's something significant like Mito Komon's seal, especially since I don't understand the reality of the *p* value. It's precisely because *p* value is an incomprehensible probability that not only the audience but also the researcher themselves might be misled into thinking they have done high-level research. While it's needless to say that PHC, which can convey the significance of research in an understandable manner, is necessary ... I must confess that, as a researcher, I'm inwardly anxious that by having my research results expressed in an understandable way, I might be more disappointed by the smallness of those results."

Discovering useful, surprising, and impactful phenomena, and ones that can be replicated, is very difficult. The experiments retried in the 2015 Science paper [3] on the reproducibility crisis were all textbook-level studies published in top journals. Yet, their reproducibility was low. The level below that can be inferred. Until now, a vast number of academically meaningless but significant articles have been published in journals, protected by the *p* value, whose reality is unknown. They were published in large quantities because rejecting the null hypothesis, a mere necessary condition for valuable research, was the criterion for peer review. Not only athletes but also humans and zombies were sent to the Olympics, keeping numerous journal editorials running. However, considering that the absolute number of excellent studies is small, if we make "the probability that a proposition contributing to the progress of the respective discipline is correct" the obvious criterion for passing peer review, some academic journals might indeed run out of articles to publish.

My Supervisor Says, "Do Significance Testing"

So, a professor with the authority to confer degrees like a bachelor's, master's, or doctorate, or the person in charge of a course, instructs you to perform significance testing. You may want to periodically ask the following questions and then ask, "Do I still need to use significance testing?"

1. The American Statistical Association issued a "Ban on Significance Testing." They explicitly forbid saying "Statistically Significant" in the imperative form. What your professor is saying is like telling you to ride in a car that the manufacturer has recalled and banned from being driven.
2. Nature announced "Let's Retire Significance Testing," with over 800 famous scientists and statisticians signing it. Your professor is like telling you to ride in a car that the largest consumer group has called for caution in using.

3. Rejecting the null hypothesis is just a necessary condition for useful scientific results. There are plenty of situations where the null hypothesis is false and yet the research is scientifically meaningless. We should switch to a methodology that probabilistically states propositions that are sufficient conditions for being scientifically useful.
4. For controlled experiments, it's desirable that multiple groups are homogeneous before the treatment. However, significance testing cannot actively support the null hypothesis. We should use a method that actively verifies that multiple groups are substantially equal.
5. The APA style manual says, if you do significance testing, show the basis for deciding n. I know three types of power analysis methods, but to freely choose a thesis topic, one needs to master power analysis for dozens of tests. Please teach me the remaining dozens of power analyses so I can comply with the manual.
6. I'll do interim tests to check the progress of the research, so I need to adjust for sequential testing and adaptive planning. It's mandatory in the APA manual. I don't want to violate the manual, so please teach me sequential testing and adaptive planning.
7. In survey research, etc., why do we do multiple comparisons only within levels when product-order α inflation occurs due to levels, attributes, reference variables, and their combinations? The results will not be reproducible. Let's use a method that doesn't create multiplicity issues.
8. If I can't prove after writing that I decided n in advance, I'll be suspected of monitoring p-values. I can't claim my innocence of not p-hacking. If someone (e.g., Mr. T) later says, "If you can't prove you decided n in advance, I'll sue to revoke your degree," how will you argue? Since the burden of proof of academic value lies with the writer, I will definitely lose if sued. So please guarantee forever, as part of your duty in degree certification, that I decided n in advance.

The answers to these questions will reveal your supervisor's academic insight.

References

1. Wasserstein, R., Schirm, A., and Lazar, N. (2019) Editorial: Moving to a World Beyond "$p < 0.05$", The American Statistician, 73: 1–19. The original text of the section cited in the main body is Chapter 2, Don't Say "Statistically Significant", line 4, "We conclude, based on our review of the articles in this special issue and the broader literature, that it is time to stop using the term "statistically significant" entirely", and Chapter 4, "Editorial, Educational and Other Institutional Practices Will Have to Change."
2. Amrhein, V., Greenland, S., McShane B, and more than 800 signatories (2019) Retire statistical significance. *Nature*, 567(7748):305–307. The original text of the section cited in the main body is "We agree, and call for the entire concept of statistical significance to be abandoned."
3. Open Science Collaboration. (2015). Estimating the reproducibility of psychological science. *Science*, 349, aac4716.
4. Wasserstein RL, Lazar NA. (2016) Editorial: The ASA's statement on p-values: Context, process, and purpose. *The American Statistician* 70: 129–133.
5. Bem, D. J. (2011). Feeling the future: Experimental evidence for anomalous retroactive influences on cognition and affect. *Journal of Personality and Social Psychology*, 100, 407–425.
6. Toyoda, H. (2002). Behavioral biases and impermanence. *Psychology World, 18*, 25–28. (In Japanese).
7. Yanagawa, T. (2017). The p-value as an excellent ruler for reporting results of clinical research data analysis. *Biometrics*, 38(2), 153–161. (In Japanese).
8. Toyoda, H. (2016). *First-time statistical data analysis: Bayesian statistics in the post-p-value era*. Asakura Publishing. (In Japanese).
9. Suzukawa, Y., and Toyoda, H. (2012). Prospective case analysis of effect size, test power, and necessary sample size in psychological research. *Psychological Research*, 83, 51–63. (In Japanese).
10. Cohen, J. (1988). *Statistical power analysis for the behavioral sciences*. 2nd ed. Hillsdale, NJ: Lawrence Erlbaum Associates.
11. American Psychological Association. (2019). *Publication Manual of the American Psychological Association (7th ed.)*. [Maeda, J., Eto, H., and Tanaka, T. (Trans.) (2011). APA Paper Writing Manual, 2nd Edition, Igaku-Shoin (Translation of the 6th edition original)].
12. Jennison, C., and Turnbull, B. W. (2000) "Group Sequential Methods with Applications to Clinical Trials," Chapman & Hall/CRC, Boca Raton.
13. Chow, S. C., and Chang, M. (2011) "Adaptive Design Methods in Clinical Trials" (Chapman & Hall/CRC Biostatistics Series).

H. Toyoda, *Statistical Significance and the PHC Curve*,
https://doi.org/10.1007/978-981-97-7748-8

14. Kruschke, J. (2014). *Doing Bayesian Data Analysis, Second Edition: A Tutorial with R, JAGS, and Stan*. Academic Press. [Maeda, K., and Kosugi, T. (Supervising Trans.) (2017). Bayesian Statistical Modeling: A Tutorial with R, JAGS, and Stan, 2nd Edition, Kyoiku Publishing. Chapter 12]. (In Japanese).
15. Toyoda, H. (Ed.) (2015). *Bayesian statistics from the basics: A practical introduction with Hamiltonian Monte Carlo*. Asakura Publishing. (In Japanese).
16. Birnbaum (1962) On the foundations of statistical inference. *Journal of the American Statistical Association*, 57, 269–326.
17. Toyoda, H. (2017). *Psychological statistics '17: Moving beyond significance testing*. The Open University of Japan. (In Japanese).
18. Toyoda, H. (2020). *Continued first-time statistical data analysis*. Asakura Publishing. (In Japanese).
19. Toyoda, H. (Ed.) (2018). *Fun Bayesian modeling: Expanding the research frontier with case studies*. Kitaoji Shobo. (In Japanese).
20. Toyoda, H. (Ed.) (2019). *Fun Bayesian modeling 2: Expanding the research frontier with case studies*. Kitaoji Shobo. (In Japanese).
21. Toyoda, H. (2024). *Statistics with Posterior Probability and PHC Curve*. Springer.

The manufacturer's authorised representative in the EU is Springer Nature Customer Service Centre GmbH, Europaplatz 3, 69115 Heidelberg, Germany. If you have any concerns regarding our products, please contact ProductSafety@springernature.com

Printed and bound by CPI Group (UK) Ltd, Croydon, CR0 4YY
01/10/2026
02225876-0016